Liz Strachan taught in her local school in Montrose, north-east Scotland, for most of her long career as a teacher of mathematics. She enjoys writing and has had two hundred articles, letters and short stories published. Liz has also won several prizes at the Scottish Association of Writers' Annual Conference.

She is a keen golfer, but admits to having no talent in her favourite sport whatsoever.

Liz is married with two sons, four grandchildren and one gorgeous great grandson.

Also by Liz Strachan

A Slice of Pi
Numbers are Forever

EASY AS Pi

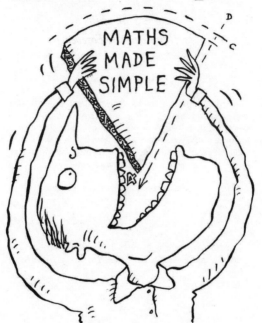

MATHS MADE SIMPLE

Liz Strachan

ROBINSON

ROBINSON

First published in Great Britain in 2016 by Robinson

Copyright © Liz Strachan 2016

1 3 5 7 9 10 8 6 4 2

A CIP catalogue record for this book
is available from the British Library.

ISBN: 978-1-47213-728-9

Typeset in Great Britain by Initial Typesetting Services, Edinburgh
Printed and bound in Great Britain by Clays Ltd, St Ives plc

Papers used by Robinson are from well-managed forests
and other responsible sources

MIX
Paper from
responsible sources
FSC® C104740

Robinson
An imprint of
Little, Brown Book Group
Carmelite House
50 Victoria Embankment
London EC4Y 0DZ

An Hachette UK Company
www.hachette.co.uk

www.littlebrown.co.uk

Acknowledgements

I would like to thank my editor, Duncan Proudfoot, for persuading me to write this book and for being patient and supportive throughout.

I would also like to thank Emily Byron, Steven Appleby for his fantastic artwork, and Nick Fawcett for his careful and thorough copyediting of the manuscript.

Contents

Introduction

When you glanced at the title of this book, you perhaps detected a spelling error. But it was deliberate. *Easy as Pi* will be exactly that – but first, let's find pi.

Collect your tape measure and any round bowl – a flower pot is ideal. Measure round the rim (mathematicians call this C, the circumference), then measure across the widest part of the bowl which is D, the diameter. With a pocket calculator or the one on your phone, divide C by D. You'll get 3 and lots of decimal places. If the first decimal place is 1 or 2, you are brilliant. You have discovered a special number which mathematicians call pi, denoting this with the Greek letter π, just to make it a bit more mysterious.

There will be more about pi later, but meanwhile let's discover that maths is fun.

The Importance of Being Zero

Zero or 0 is an important number. We use it all the time in our calculations. Yet the Greeks did all that important geometry without it and the Romans managed very well with M, D, C, L, X, V and I. Europe was without a zero until the Middle Ages. It took many centuries to develop and many countries to cross. But when it arrived at last via the Italian mathematician Fibonacci, in his book *Liber Abaci* (The Book of Calculation), it was a huge moment in the history of mathematics.

Zero, of course, is a number in its own right although it is also used as a place-holder, making the difference between 35 and 3005 very clear.

Some say that zero is neither even nor odd. However, the definition of an even number is that it can be divided by 2 without leaving a remainder: $0 \div 2 = 0$ and nothing can be left over if there was nothing there in the first place. So, zero is definitely even.

Add, Subtract, Multiply and Divide with Zero

We will use a touch of algebra here. The curly x is called a variable, that is, its value can vary and, in this case, it stands for 'any number'.

(In algebra, you must always use the curly x to make it different from x, the sign for multiply. Otherwise, in a young student's handwriting, $x \times x$ would look like xxx, like three kisses on a birthday card.)

$$x + 0 = x$$
$$x - 0 = x$$
$$x \times 0 = 0$$
$$x \div 0 = ?$$

The first three equations above are straightforward but you don't know the answer to the last one, do you? Well, join the club! Nobody in the whole world knows! Try it on your calculator. It either shouts 'ERROR!' or simply says 'undefined'. There is a saying that black holes appeared when God tried to divide a number by 0.

But, an empty box of chocolates divided between x friends will give zero satisfaction so $0 \div x = 0$. However, $0 \div 0$ is also impossible and undefined.

Zero in Sport

There are a few strange words used in sport for 0.

In cricket, if you are 'out for a duck', you walk off the pitch, head down, humiliated, embarrassed, because you haven't even managed to score a single run. Oh, the shame of it!

In tennis and other racquet sports 'love' means 'nothing', as many adoring fans know to their sorrow. Some think it comes from the French word *oeuf* meaning 'egg', a 0 being egg-shaped. However, the pronunciation of love and *oeuf* are not similar and anyway, in France, the term is just *zéro* so the 'egg' explanation is a bit far-fetched.

In football, 0 is always called 'nil'. So the unlikely score on *Sports Report* might be 'Scotland three, England nil'. (Well, one can dream!)

In golf, if you play off 'scratch', your skill at the game is simply staggering. You have a zero handicap so if you play in a competition against a poor golfer with a handicap of 36, you have to give them two strokes at every hole. Thus if you have a 4 at the first hole and your opponent has a 6, the hole is halved. This only happens in club golf in order to give poor golfers a sporting chance.

A Stern Reminder

In maths, 0 is always called zero. It is not the done thing to call this important number nothing, nil, nada or zilch.

A million is one with six zeros. It is not 'one, oh, oh, oh, oh, oh, oh', like a reaction to some medieval torture.

Zero tolerance will be imposed and breaking the rule will result in writing out the 13 times table three times.

And leave 'nought' to the poets:

> The best-laid schemes o' mice an' men
> Gang aft agley,
> An' lea'e us nought but grief an' pain,
> For promis'd joy!
>
> *Robert Burns*

Adding

Adding is commutative. That's a maths word you probably don't know the meaning of. You will find that mathematicians like to increase the mystery of their subject by using long, obscure words.

It simply means that numbers to be added can change places. Take, for example, 7 + 28 + 13. You would be happier with 7 + 13 + 28 because 7 and 13 are pals. You're then able to work out 48 immediately. You can do something similar with 56 + 53 + 47. This time, add 53 and 47 first, then add 56 to make 156.

Your number brain won't like 9 + 276. It prefers the sum the other way round. It is even easier to calculate 276 + 10 − 1.

Adding Consecutive Numbers

A little maths genius, named Carl Friedrich Gauss, was born in Brunswick, Germany, in 1777, the only son of poor peasants. His parents scrimped and saved and managed to pay for some elementary education, and so began the career of one of the greatest mathematicians of all time.

When Carl was only seven, he and his classmates were instructed to add up all the numbers from 1 to 100. Their teacher,

Herr Buttner, thought this huge addition sum would keep the class quietly busy for a long time and he could have a rest. But after only a few seconds, Carl raised his hand and said, 'The answer, sir, is 5050.'

The boy had noticed that the numbers 1 to 100 were in fifty pairs: 1 and 100, 2 and 99, 3 and 98, 4 and 97, 5 and 96, and so on. Each pair added to 101 and since there were fifty pairs, the answer was 50×101 or 5050. Or another way, Carl pointed out, was simply to write down the highest number, 100, multiply by the next number after that, 101, and divide by 2.

And, Carl added, if Herr Buttner wanted to know the sum of only the odd numbers from 1 to 100, that was equally easy. Simply divide 100 by 2, and square it, getting $50 \times 50 = 2500$. The sum of the even numbers would therefore be $5050 - 2500 = 2550$. Test this for yourself.

What is the sum of all the numbers from 1 to 10?

$$1 + 2 + 3 + 4 + 5 + 6 + 7 + 8 + 9 + 10 = 55$$

Or, by the Gauss method:

$$10 \times 11 \div 2 = 55$$

What is the sum of all the odd numbers from 1 to 10?

$$1 + 3 + 5 + 7 + 9 = 25$$

Or, by the Gauss method:

$$10 \div 2 \times 5 = 25$$

The sum of the even numbers is $2 + 4 + 6 + 8 + 10 = 30$ or $55 - 25 = 30$.

Our little brain of Germany would not have been popular with his teacher that afternoon, but Herr Buttner recognised the boy's genius and arranged for sponsors to pay for all his future education.

Carl went on to become one of the world's greatest mathematicians. He was such a genius that, after his death, his brain was removed to see if it was any different from number-numpty brains or even ordinary brains – but it wasn't.

Averages

There are three different ways of calculating an average. They are called the mean, median and mode. One method is usually more appropriate than the other two.

Mean

June in the north-east of Scotland was unusually cold last year. No one was seen on the golf course wearing shorts and no one needed Factor 50 sun cream. Throughout the month, the daily temperature at noon was in degrees Celsius:

12 13 15 14 13 15 14 16 16 16 13 14 13 12 16
15 17 17 17 14 16 16 15 14 13 15 17 16 18 18

To calculate the mean, add the thirty numbers and then divide by 30. Adding numbers in a row makes you cross-eyed. You could use your calculator, but it would be better if you had a friend to call out the numbers as you press the buttons.

However, since all the numbers are quite similar, make it easier for yourself. Look for the lowest number, which is 12, and subtract 12 from each measurement. Rewrite the numbers:

0 1 3 2 1 3 2 4 4 4 1 2 1 0 4
3 5 5 5 2 4 4 3 2 1 3 5 4 6 6

This is easier to add. The top row is 32 and the second row is 58, a total of 90. The mean is therefore 90 ÷ 30 = 3, but don't forget to add on the 12 you subtracted earlier. The mean temperature for June is therefore 15°C. As they say here, 'Chilly for June.'

The following two types of average do not involve adding but need to be included in this section.

Median

If a class of students is a homogeneous bunch, finding the average mark for their end-of-term maths test would probably be done by calculating the mean as above.

However, one of the students is a twenty-first-century Carl Friedrich Gauss and always gets 100 per cent in every test. There are also two new students who ought to have been in a lower class and have failed the exam, and Lazy Larry who hasn't done a stroke of work all term.

The class of twenty have percentage results as follows:

86 67 71 100 74 68 85 66 15 18 73 66 75
30 69 81 77 82 68 74

Now, including Brainy Brian's 100 per cent is fair enough. It will increase the mean average and impress the headmaster, but the three bad scores will lower the class average unfairly.

So the best way to calculate the average this time is by finding the median or the middle score.

Arranging the results in order, we get:

15 18 30 66 66 67 68 68 69 71 : 73 74
74 75 77 81 82 85 86 100

So the median average lies between 71 and 73 and is therefore 72.

Another Use for the Median

The best way to find the average height of, say, a rugby team is to line them all up in order, shortest first, tallest last and all the others in their correct place, and pick out the median average, the eighth one in the middle. Then ask him how tall he is.

Mode

A newspaper reported recently that the average dress size for women was now 16. Complicated calculations were not necessary here. The only data needed was the 16, the size that was selling best. This is called the **mode**. Sometimes the average is bimodal where two sizes are equally popular.

The First of Several Terrible Maths Jokes

In the entire history of mathematics, there has never once been a maths joke that would make any normal person laugh out loud or even smile. However, the brainier and nerdier the mathematician, the more they enjoy a maths joke. Warning! If you find any of the jokes remotely funny as you progress through this book, you may be in the first stage of becoming a nerdy maths egghead.

Joke: A mathematician drowned when he tried to wade across a river known to have an average depth of four feet.

Eight Eights

Now, after all that adding, here's a bit of light relief. What do eight 8s make? 64? Not this time. We've been adding, remember!

$$888 + 88 + 8 + 8 + 8 = 1000$$

A Little More Adding

Here is a chance to show off how good your adding is. In fact, you will be able to add four numbers correctly before you even know what the numbers are!

You are going to be like Nicomachus, the first-century Jordanian mathematician who loved to waylay his fellow citizens in the street and do some number work with them. But, to avoid arrest, it would be best if you choose your victims from within the family.

Draw a grid like this and write **34** on a piece of paper, fold it and ask your 'volunteer' to put it in their pocket.

1	2	3	4
5	6	7	8
9	10	11	12
13	14	15	16

Now, ask them to choose a number – let's say 7 – and then shade out the remaining numbers in that row and column.

1	2	3	4
5	6	7	8
9	10	11	12
13	14	15	16

Next, ask them to choose two more numbers. They choose 2 and 13 and shade out the other numbers in these rows and columns.

1	2	3	4
5	6	7	8
9	10	11	12
13	14	15	16

Finally, they will see there is only one other square not shaded out and that is 12, so this has to be their last choice.

Ask them to add 13 + 2 + 7 + 12 (= 34) and they will be amazed that it's the same number as the one in their pocket. They may

Calendar /

Outline any 3 × 3 square block (
any year of a calendar.

Here is July 2016:

8 15

9 16

10 17

The total will always be nine time
the square. In this case:

16 × 9 =

Outline any 5 by 4 block:

2   3   4

9   10 11

16 17 18

23 24 25

To find the total easily, add the
and multiply by 10. In this case:

2 + 27 = 29 ×

want to repeat the trick but don't do it! You see, for a 4 × 4 grid the answer is always 34, so they might become suspicious.

Ask them to draw a larger grid, 5 × 5, 6 × 6 or even larger, and repeat all the instructions. For example, for the 5 × 5 grid get them to choose and circle five numbers. The total this time will be 65 and for the 6 × 6 grid, it will be 111.

Where does the magic total come from? Add up the numbers in the diagonal and there's the answer.

More Easy Adding

Here is another adding trick to play with a young friend. The first chosen number can be three, four, five or more digits, but three will be perfect to start with. You (Y) will invite a friend (F) to choose a three-digit number. (Find some excuse to dissuade F from choosing 999.) In fact, be generous. Ask F to write down two numbers as follows:

427

635

Memorise 427.

Now it is your turn so you pick 364. It has to be 364 because you are deliberately making the second number up to 999. F now chooses the fourth number, say 128, and again you will make F's number up to 999 by choosing 871.

Now you have five number

4
6
3
1
8

Ask F to add t

While he is doing this painstaki
tion sum and announce that you
head and the total is 2425. F wil
are a genius. But all you did wa
to the front of it and subtract 2

Try it again for a five-digit

(F)
(F)
(Y)
note that lines 2 ar
(F)
(Y)
again note th

Now you do as before. Attach
the last 3, getting your total o
and marvels at your newly disc

Subtraction

Subtraction is not commutative. But you can fiddle about with subtraction in a way you couldn't do with adding. For example:

$$32 - 13 = 19$$
Take 3 from the 32 and 13, getting
$$29 - 10 = 19$$
Add 7 to the original 32 and 13, getting
$$39 - 20 = 19$$
Take 12 from the original 32 and 13, getting
$$20 - 1 = 19$$

Get the idea? So, if $86 - 19$ doesn't appeal to you, add 1 to both numbers, getting $87 - 20 = 67$. Your brain likes this much better.

Most people don't like subtracting from numbers with lots of zeros, for example $1000 - 847$. Simple! Just take 1 from each number, getting

$$\begin{array}{r} 999 \\ -\ 846 \\ \hline 153 \end{array}$$

What about 104 – 67? You'd like 99, wouldn't you? Well, you know what to do.

$$\begin{array}{r} 99 \\ -62 \\ \hline 37 \end{array}$$

So, in subtraction, life is much easier when the number in the top line has each digit larger than the corresponding digits in the bottom line; for example:

$$\begin{array}{r} 8795 \\ -2164 \\ \hline 6631 \end{array}$$

But that doesn't often happen and you are probably remembering that subtraction sums usually involved the dreaded 'borrowing'.

Wait though! Sometimes, you can avoid it.

$$\begin{array}{r} 752 \\ -418 \\ \hline \end{array}$$

Oh dear, we can't do 2 – 8. But we can do 52 – 18 (or, remembering what you did above, 54 –20), getting 34, so the answer without the need to borrow is 334.

Another similar example:

$$854$$
$$-247$$

We can't do 4 – 7 but we can do 54 – 47 getting 7, or 07, so the answer is 607.

Even a more difficult example like 6308 – 1279 can be tackled without borrowing. 308 – 279 (or 309 – 280) = 29, or 029, so the answer is 5029.

But, sometimes, maybe, you might just *have* to use the old borrowing method:

$$834$$
$$-267$$
$$\overline{567}$$

Remember the old way: 7 from 4, you cannot, so borrow 1 from the 3 making 4 into 14 and reducing the 3 to 2. So 7 from 14 = 7. Write down the 7. Now 6 from 2, you cannot, so borrow 1 from the 8 making the 2 into 12 and reducing the 8 to 7. So 6 from 12 = 6. Write down the 6. And finally, 2 from 7 = 5. So the answer is 567. Whew! That was hard work!

And if, by any chance, you are wondering why the 4 became 14 and the 2 became 12, it is because of something you did in your dim and distant past. Remember? Above the top line, you used to write H T U (hundreds, tens and units) so the 1 borrowed from the 3 is actually worth 10.

SWOON. Wake me up when we get to page 30.

Poor soul, if you need a lie-down after that, you can forget all about this method and do what we did before.

$$834$$
$$\underline{267}$$

Change 834 to a number you like better such as 799. Subtract 35 from the top and bottom lines and you have 799 − 232 = 567.

As Polonius said to his hot-headed son, Laertes, 'Neither a borrower nor a lender be' (*Hamlet* by William Shakespeare).

The 6174 Trick

Here is an easy trick to practise your subtraction. Choose any four-digit number (except one with all the same digits.) For example, 7623. Now rearrange the digits to make the highest number and rearrange them again to make the smallest number and subtract them. Use whichever subtraction method is easiest.

$$7632$$
$$-2367$$
$$5265$$

Repeat this process:

$$6552$$
$$-2556$$
$$3996$$

Repeat:

$$9963$$
$$-3699$$
$$6264$$

Repeat:

$$6642$$
$$-2466$$
$$4176$$

Repeat:

$$7641$$
$$-1467$$
$$\mathbf{6174}$$

And that's it! After that you will have 6174 forever.

If you really have nothing better to do, try one yourself. I can guarantee you will end up with 6174 but I can't foretell how many 'repeats' you will have to get there.

I add and subtract using my fingers.

Easy peasy!

Multiplying

Perhaps you were unlucky. You might not have had a primary teacher like mine. Miss Drummond made us chant the times tables every morning. That, of course, was after the Lord's Prayer – 'Our Father, who chats in heaven, Harold be his name . . .' I had an Uncle Harold in Australia but he sent me exciting post-war Christmas presents, so he obviously wasn't in heaven.

I knew my times tables so well that I could have been wakened from a deep sleep and given the answer to 12 × 9 before I remembered my name.

Drummie was a strict disciplinarian. She believed in the old saying, 'Give them an inch and they'll take a mile.' There were forty of us and she ruled with a rod of iron. The punishment for the slightest misdemeanour was writing out a times table over 12. I am proud to say I know my 19 times table and all the ones below that.

Don't you mean, Miss, that if you give me a centimetre I'll take a kilometre?

She was a firm believer in chanting. There was more chanting in our classroom than in a Buddhist monastery. We chanted the books of the Bible (Genesis, Exodus, Leviticus, Numbers, Deuteronomy, Joshua, Judges, Ruth . . .) and our history ('In 1492, Columbus sailed the ocean blue') and we also chanted all the other tables needed in our maths lessons:

16 ounces, 1 POUND
14 pounds, 1 STONE
8 stones, 1 HUNDREDWEIGHT
20 hundredweights, 1 TON.

Now, although these draconian days are over, there still is the need to learn the times tables by heart. The times tables are the building blocks for maths to come later, like fractions and algebra.

Students who have not memorised their tables will find future maths topics more difficult than they need to be.

OK, there are calculators and everybody has one, the handiest being on your mobile phone. They are great tools for long, laborious calculations but it is unnecessary to use your calculator for the £62.90 lunch bill for six. Round it up to £66 or £72 because you know these are 6 times table numbers. That's £11 each if the service is bad or £12 each when the food is good.

Students who rely on calculators are usually weak at estimating skills and are unaware of wrong answers caused by pressing the wrong buttons. 'Garbage in, garbage out' (GIGO) was first said by Charles Babbage, the inventor of the Mechanical Difference Engine, the forerunner of the modern electronic computer.

So get these tables into your head. Repeat them, chant them, sing them, play multiplication games. It is worth the effort. Confidence with the times tables prevents anxiety, stress and embarrassment.

Facts about the Tables

Three

The **3** times table doesn't need to stop at $3 \times 12 = 36$. It can go on forever:

$$3 \times 37 = 111, 3 \times 158 = 474, 3 \times 4582 = 13\,746.$$

Add the digits in each answer and you get 3, 15, 21, all of them on the 3 times table. Is the number 1423 divisible by 3?

$$1 + 4 + 2 + 3 = 10, \text{ so } 1423 \text{ is not divisible by 3.}$$

Four

The **4** times table beyond 48 includes 104, 716 and 1128. For divisibility you don't need to check the entire number. Just look at the number formed by the last two digits. If the number is divisible by 4, then so is the entire number. Is 34 514 divisible by 4? (No.)

Five

Multiplying by **five** is easy. But can you, for example, do 13 684 × 5 without paper and pencil? Don't tell me you were thinking of using your calculator! It's simple! The answer is 68 420. Without multiplying and all that 'carrying' stuff, just divide by 2 and attach **0**. Confirm these: 468 × 5 = 2340, 6078 × 5 = 30 390 and 81 244 × 5 = 406 220.

It's just as easy if the last digit is odd. Just ignore the **remainder** and attach **5** instead of 0. Confirm these: 667 × 5 = 3335, 1295 × 5 = 6475 and 46 343 × 5 = 231 715.

Six

The **6** times table behaves like the 3 times table but for a number to be divisible by 6, the last digit has to be **even** and the total of the digits has to be divisible by **3**. For example, 1821 divides by 3 but it is not even, so it is not divisible by 6. The digit total for 156 divides by 3 and it is even, so it definitely divides by 6. And, at a glance, you can see that 1 000 002 is divisible by 6.

Seven

The **7** times table is nobody's favourite. In fact, it's a nasty little table. If, suddenly, 7 × 6 eludes you, swap it round and do 6 × 7 instead.

There is no way of telling at a glance if a number is divisible by 7, but the following method is simple. This is what you do: double the last digit of the number and subtract it from the rest of the number. Keep doing this until you get (or don't get) 0 or an obvious 7 times table number.

Examples: Is 7203 divisible by 7? Well, 720 – 6 = 714 and 714 obviously divides by 7 so 7203 is divisible by 7.

What about 4102? 410 – 4 = 406 and 40 – 12 = 28, which is on the 7 times table, so 4102 must be divisible by 7.
And 15 633? 1563 – 6 = 1557 and 155 – 14 = 141 and 14 – 2 = 12. So 15 633 is not divisible by 7.

At Sixes and Sevens

$$6 \times 7 = 42$$
$$66 \times 67 = 4422$$
$$666 \times 667 = 444\,222$$
$$6666 \times 6667 = 44\,442\,222$$
$$66\,666 \times 66\,667 = 4\,444\,422\,222$$

Guess what 666 666 × 666 667 would be?

Eight

The **8** times table goes 8, 16, 24, 32, 40, 48, 56, 64, 72, 80, 88, 96. In order to tell at a glance if a number is divisible by 8, the number formed by the last three digits of the number must divide by 8. So 320 192 divides by 8 because 192 ÷ 8 = 24.

A Little Flaw

123 456 789 × 8 = 987 654 312 (not quite perfect, is it?)

But this is definitely perfect!

$$1 \times 8 + 1 = 9$$
$$12 \times 8 + 2 = 98$$
$$123 \times 8 + 3 = 987$$
$$1234 \times 8 + 4 = 9876$$
$$12\,345 \times 8 + 5 = 98\,765$$
$$123\,456 \times 8 + 6 = 987\,654$$
$$1\,234\,567 \times 8 + 7 = 9\,876\,543$$
$$12\,345\,678 \times 8 + 8 = 98\,765\,432$$
$$\mathbf{123\,456\,789 \times 8 + 9 = 987\,654\,321}$$

A Really Pathetic Maths Joke

Why did 6 fear 7? Because 7 8 9.

Nine

You can't go wrong with the **9** times table: 9, 18, 27, 36, 45, 54, 63, 72, 81, 90, 99, 108. The digits in each number add to 9. And that is the rule for divisibility by nine. Is 49 338 divisible by 9? It must be, because the digits add up to 27.

Here is a magic pattern of numbers using the 9 times table:

$$12\,345\,679 \times 9 = 111\,111\,111$$
$$12\,345\,679 \times 18 = 222\,222\,222$$
$$12\,345\,679 \times 27 = 333\,333\,333$$
$$12\,345\,679 \times 36 = 444\,444\,444$$
$$12\,345\,679 \times 45 = 555\,555\,555$$
$$12\,345\,679 \times 54 = 666\,666\,666$$
$$12\,345\,679 \times 63 = 777\,777\,777$$
$$12\,345\,679 \times 72 = 888\,888\,888$$
$$12\,345\,679 \times 81 = 999\,999\,999$$

You will notice that 8 has been banned from this pattern because it just messes things up.

However let's include 8:

$$123\,456\,789 \times 9 = 1\,111\,111\,101$$
$$123\,456\,789 \times 18 = 2\,222\,222\,202$$
$$123\,456\,789 \times 27 = 3\,333\,333\,303$$
$$123\,456\,789 \times 36 = 4\,444\,444\,404$$

And so on.

Quite a spectacular pattern too, isn't it?

Eleven

The **11** times table is easy until you come to 11 × <u>11</u> = 121 and 11 × <u>12</u> = 132, but look at the underlined 11 and 12. In both cases, leave a space between the digits, 1 1 and 1 2, and insert the sum of the two numbers between them, getting 121 and 132.

In the same way, we can do: 13 × 11 = 1**4**3, 14 × 11 = 1**5**4, 15 × 11 = 1**6**5, then 21 × 11 = 2**3**1, 22 × 11 = 2**4**2 and dozens of other two-digit numbers. For 67 × 11 (and others where the digits add to 10 or more), write down the digits as before, 6 7, but 6 + 7 = 13 so you have to carry 1, changing the answer to 737.

An Easy Way to Multiply Any Number By 11

By the following method, multiplying even a large number by 11 only takes a few seconds. For example: 724 361 × 11.

Write down the first and last digits with a space in between:

7   1

Now add the digits in pairs from right to left: 1 + 6 = 7, 6 + 3 = 9, 3 + 4 = 7, 4 + 2 = 6 and 2 + 7 = 9. Now fill in 96797 in the space, getting the answer 7 967 971. (You have to work from right to left because if the pair of digits adds to 10 or more, for example 12, you have to write down the 2 and carry 1.

Example: 269 × 11.

Write down 2 and 9 leaving a space between them. 6 + 9 = 15 so write 5 and carry 1; 2 + 6 + the carried 1 = 9, so the answer is 2959.

Before we go on to our last times table, here is a pretty pattern to show you the magic of maths:

A Number Pattern for Legs Eleven and Friends

$$11 \times 11 = 121$$
$$111 \times 111 = 12\,321$$
$$1111 \times 1111 = 1\,234\,321$$
$$11\,111 \times 11\,111 = 123\,454\,321$$

And up and up you go to 9-legs × 9-legs:

$$111\,111\,111 \times 111\,111\,111 = 12\,345\,678\,987\,654\,321$$

It doesn't work after that because 10-legs × 10-legs would have 10 in the middle of the answer, necessitating a carried 1, and the pattern becomes very messy. But 13 legs × 13 legs is nicely symmetrical and equals: 12 345 678 900 987 654 321.

I'm Legs Eleven!

30

Twelve

Lastly, the **12** times table. An easy way to learn it is to write down the 10 times table in a column, along with the 2 times table, and add them:

$$10 + 2$$
$$20 + 4$$
$$30 + 6$$
$$40 + 8$$
$$50 + 10$$
$$60 + 12$$
$$70 + 14$$
$$80 + 16$$
$$90 + 18$$
$$100 + 20$$
$$110 + 22$$
$$120 + 24$$

So the numbers for the 12 times table are: 12, 24, 36, 48, 60, 72, 84, 96, 108, 120, 132, 144.

Long Multiplication

I can almost hear you sighing in deep despair. However, fear not! Even competent mathematicians use a calculator to do long multiplication. Not to do so would be like living with candlelight long after electricity had been discovered or washing the sheets in the bathtub when there is a washing machine in the kitchen.

So feel free to use your calculator if by chance you need to multiply, for example, 194 by 378.

But! You should have a rough idea of what the answer will be. So 194 × 378 is approximately 200 × 400 = 80 000 and your calculator answer will be a little less than that. Also 194 and 378 have last digits 4 and 8 and 4 × 8 = 32, so the last digit in your answer will be 2. So go ahead and key in your numbers and you get 73 332. The answer to the multiplication is less than 80 000 and the last digit is 2 so you can safely assume that your answer is correct.

By the way, instead of saying, 'What is the answer when you multiply 194 by 378? ' get into maths mode and say 'What is the **product** of 194 and 378?' In maths, product means the result of multiplying.

Shorthand Signs You Should Know

< means 'less than'
> means 'more than'
≤ means 'less than or equal to'
≥ means 'more than or equal to'

For example, 3 < 6 and 7 > 5. Note that the pointer always points to the smaller number. 'Choose a number $x \geq 10$' means any number from 10 upwards.

≠ means 'not equal to'

≈ means 'approximately equal to'

( ) brackets mean that the numbers inside should be done first;
for example, $5 \times (6 + 2) = 5 \times 8 = 40$.

Long Division

Drummie's chant for long division was 'into, multiply, subtract, bring down – into, multiply, subtract, bring down . . .'

Let's try it but remember there *might* be the calculator to fall back on, if long division proves to be the straw that broke the camel's back.

OK, Drummie, here we go. 51 **into** 7, can't do, 51 into 74 is 1 (write 1) now **multiply** 1 by 51 and **subtract**. Now **bring down** 8. Repeat the four instructions.

$$
\begin{array}{r}
14 \\
51{\overline{)748}} \\
-51 \\
\hline
238 \\
-204 \\
\hline
34 \\
\end{array}
$$

Not too bad, was it?

To make long division even more daunting, there is some scary maths vocabulary to go with it. The 14 at the top is called the **quotient**, the 748 is the **dividend**, 51 is the **divisor** and, 34, the bit left at the end, is the **remainder** (R).

But it is worth persevering. Try a few more yourself:

    (a) 3572 ÷ 16
    (b) 9856 ÷ 31
    (c) 4668 ÷ 39
    (d) 67 983 ÷ 145

Answers: (a) 223 R4 (b) 317 R29 (c)119 R27 (d) 468 R123

However, if you are now a very unhappy bunny, you *can* use a calculator, even when there is a remainder.

AN UNHAPPY BUNNY

Let's take (a) above. Key in 3572 ÷ 16 and you get 223.25. **But the remainder is not 25!** This is the decimal part of the answer and to find the actual remainder, you need to multiply 0.25 by the divisor, 16, getting, hey presto, 4. So using the calculator was easy.

However, for (b) 9856 ÷ 31 your eight-digit calculator will

read 317.93548 and when you multiply 0.93548 by 31 you only get 28.99988, but that is nearly 29 so you have your answer almost as easily as the last one. The 0.93548 is by no means an exact decimal. In fact the digits go on forever and ever and never stop. The most powerful supercomputer in the world, the Chinese Tianhe 2, would provide you with several million decimal places but there will never be a last digit.

The third example, 4668 ÷ 39, gives 119.69231, but again these digits will go on forever so 0.69231 × 39 = 26.9997 which is nearly 27.

The last one, 67 983 ÷ 145, on the calculator is 468.84827; multiplying the decimal part by 145 gives 122.99915, which is nearly 123 (or 122.999215 ≈ 123, using the symbol you were introduced to earlier). So the answer is 468 R123.

Something You May Have Forgotten

What are the answers to the following calculations?

(a) 10 + 2 × 5
(b) 26 – 10 ÷ 2
(c) 5 × 12 + 10 ÷ 5
(d) 4 × (3 + 6)

You may be puzzled when I tell you the answers for (a) (b) and (c) are 20, 21 and 62, although you probably are correct with your answer to (d), which is 36.

You see, number operators have a pecking order: × and ÷ are equally important and always come before + and –. So for (a) above, 2 × 5 must be done first. That makes 10 + 10 = 20. For (b) 10 ÷ 2 is done first so we have 26 – 5 = 21. In (c) 5 × 12 and 10 ÷ 5 are calculated first, getting 60 + 2 = 62.

But, for examples with + and – only, the numbers can be arranged in any order. So for 4 – 6 + 10, you might prefer 4 + 10 – 6. And for 60 × 16 ÷ 8, multiply and divide have the same status so it is simpler to tackle the 16 ÷ 8 first, getting 60 × 2 = 120.

A Mathematical Anagram

I bet you are a whizz kid at anagrams. Can you make a meaningful anagram out of ELEVEN PLUS TWO = ? The answer appears at the end of the next topic.

1 + 1 = 10

No, the heading is quite correct. But for clarity, it should be $1_2 + 1_2 = 10_2$ because you are about to learn about binary numbers.

Binary numbers are not used in everyday calculation. They lie deep in the innards of all electronic calculators and other computer-based devices where no one can see them. They look like this:

$$1000110111001010100001$$
$$0111001000100111101000$$
$$1110011000101010111011$$
$$1010100011100110001101$$

The average person is unconcerned about these mysterious numbers. As long as the computer is happy to accept data in our decimal number system, do its secret stuff, then churn the data back out again in decimal number form, that's all we want to know. However, *you* want to know, don't you?

Like many mathematical discoveries, binary numbers were invented long before scientists found any practical use for them.

The binary system, as the name suggests, uses two digits only: 1 and 0. Binary numbers were invented by Gottfried Wilhelm Leibniz, a genius philosopher and mathematician who lived in Germany between the mid-seventeenth and early eighteenth centuries.

The binary system is easy and enjoyable to learn and this is how it works.

Try to remember how you started to count. There were headings:

$$\leftarrow \quad \leftarrow \quad 1000s \quad 100s \quad 10s \quad Units$$

or

$$10^3 \quad 10^2 \quad 10^1 \quad 10^0$$

because they are decimal or base ten numbers. So 1742 or

$$1 \quad 7 \quad 4 \quad 2$$

means $1 \times 1000 = 1000$, $7 \times 100 = 700$, $4 \times 10 = 40$ and $2 \times 1 = 2$, giving 1742.

The headings for binary numbers are:

$$\leftarrow \quad \leftarrow \quad 32\text{s} \quad 16\text{s} \quad 8\text{s} \quad 4\text{s} \quad 2\text{s} \quad 1\text{s}$$

or

$$2^5 \quad 2^4 \quad 2^3 \quad 2^2 \quad 2^1 \quad 2^0$$

because they are binary or base two numbers.

So take the number 15, for example. This is $8 + 4 + 2 + 1$ so *one* lot of each of these goes under the appropriate headings, giving the binary number 1111_2. The subscript '2' must always be attached to make it clear that this is not a decimal number.

Now take 22. This is $16 + 4 + 2$ and putting these under the appropriate headings, remembering to fill in the missing headings with 0, gives the binary number 10110_2.

The headings, of course, can be extended as far as you need by doubling the previous heading.

Binary numbers get very long very quickly. Just for fun, let's do a monster. What about 700_{10}? The headings have to be:

512s	256s	128s	64s	32s	16s	8s	4s	2s	1s
1	0	1	0	1	1	1	1	0	0

So $700_{10} = 1010111100_2$.

Changing back from binary to decimal is even easier. 1110001_2 (using the headings) gives 1×64, 1×32, 1×16, 0×8, 0×4, 0×2, $1 \times 1 = 113_{10}$.

Leibniz must be up there in mathematical heaven bursting with pride.

PS: The misleading title of this section was based on this Terrible Maths Joke:

There are 10 kinds of people: those who understand binary numbers and those who do not.

Anagram answer:

ELEVEN PLUS TWO = TWELVE PLUS ONE

Square Numbers

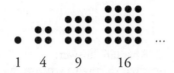

1   4   9   16

These are the first four perfect squares. The first number is obviously not a square. It is just a miserable little dot but $1 = 1 \times 1 = 1^2$ and therefore 1 has a good claim to be included. The squares go on and on forever. They are written 1^2, 2^2, 3^2, 4^2, 5^2, 6^2, 7^2 . . . and the square of an even number is even while the square of an odd number is odd, for example $12^2 = 144$ and $13^2 = 169$. A square like 57^2 is easy to do on your calculator. Just key in 57 and press the x^2 button.

But any square ending in 5, like 45^2, 95^2 or 9995^2 can be done in your head in seconds. For 35^2, just take the 3, multiply it by the next number after 3 (which is 4) and attach 25, getting 1225. Likewise $95^2 = 9 \times 10 = 90$, attach 25, getting 9025.

And $9995^2 = 999 \times 1000 = 999\,000$, attach 25, getting 99 900 025.

Your friends will be amazed at that one, especially if you say, 'Mmm, that's ninety-nine million, nine hundred thousand and twenty-five.'

The Difference of Two Squares

$$X^2 - Y^2 = (X - Y)(X + Y)$$

This is a handy little formula. It sometimes makes a calculation much easier. For example, $99^2 - 98^2$. It would be tedious to tackle this without a calculator. So, having pressed 99 and the x^2 button, then repeated this for 98, you get 9801 and 9604. Then you have to subtract – may as well use the calculator – and you finally get 197. But using the above formula, you have: $(99 - 98)(99 + 98)$, which, at a glance, you can see is $1 \times 197 = 197$.

What a pity there's no similar formula for adding two squares.

Try the following without a calculator:

(a) $243^2 - 237^2$ (d) $82^2 - 62^2$
(b) $184^2 - 176^2$ (e) $61^2 - 29^2$
(c) $149^2 - 139^2$

Answers: There's a little bit of maths magic here. All the answers are 2880.

This is Not a Rule

$$12^2 + 33^2 = 1233$$
$$\text{and}$$
$$88^2 + 33^2 = 8833$$

There aren't any other combinations of numbers that exhibit this pattern.

Every Square Number is the Sum of Two or More Consecutive Odd Numbers

$$1 = 1$$
$$4 = 1 + 3$$
$$9 = 1 + 3 + 5$$
$$16 = 1 + 3 + 5 + 7$$
$$25 = 1 + 3 + 5 + 7 + 9$$
$$36 = 1 + 3 + 5 + 7 + 9 + 11$$
$$49 = 1 + 3 + 5 + 7 + 9 + 11 + 13$$

And on and on through the never-ending square numbers and odd numbers.

A Terrible Maths Joke

I had eight tasks for my odd-job man to do. But he only did jobs 1, 3, 5 and 7.

The Tables of Imperial and Metric Length

In the US they continue to use the imperial units for measuring length. The metric system is used everywhere in Europe – except in the UK, of course, where we can't make up our minds which system we prefer and we cling on to both with great tenacity.

Most of us have little difficulty muddling along with both systems. We buy 20 square metres of carpet for a floor that has always been 15 feet by 12 feet. And we run the 26-mile marathon but do the 100-metres sprint. We know that a 6-foot man is tall, but is someone who measures 1 metre and 65 centimetres short or tall?

Our two tables of length are as follows:

Imperial Length

12 inches = 1 foot
3 feet = 1 yard
1760 yards = 1 mile

There seems to be a gap in the middle, doesn't there? That's because long ago, when I was a wee girl chanting my tables, we had three extra bits:

22 yards = 1 chain
10 chains = 1 furlong
8 furlongs = 1 mile

The furlong originated in medieval times when a furlong was the length of a furrow in a ploughed field and was 220 yards long. Furlongs are still used in the world of horse racing.

Although no one ever says, 'I'll nip down to the corner shop, it's only a few chains away', the chain lives on in some places. In cricket, the distance between the stumps is still 22 yards or one chain.

In North America the chain is used to measure the rate of spread of wildfire in chains per hour.

Metric Length

The metric table of length is much easier to learn and use:

10 millimetres (mm) = 1 centimetre (cm)
100 centimetres = 1 metre (m)
1000 metres = 1 kilometre (km)

For everyday use, 1 inch ≈ 2½ cm and 1 yard is about 3 inches shorter than a metre.

Oh, and at sea the depth is measured in fathoms and 1 fathom is 6 feet.

In the Pacific Ocean there is a very deep part called the Mariana Trench, which is nearly 6000 fathoms deep. That's nearly 7 miles!

Metric Weight

The table of metric weight couldn't be easier. It's got only two lines:

$$1000 \text{ grams (g)} = 1 \text{ kilogram (kg)}$$
$$1000 \text{ kg} = 1 \text{ tonne}$$

At the beginning of the new millennium in the UK, the Government Department of Trade and Industry announced that all food should be weighed in metric units. This decree arrived with all the clamour of a bat squeak. There were a few anxious phone calls to early morning radio programmes, but once it was confirmed that prices wouldn't go up, nobody cared or bothered to use it.

Shopping has gone on as usual; only the dialogue has changed. A purchase at the butcher shop might now have the following script:

Customer: 'A pound of mince, please.'

Butcher: 'That's a little bit under the half kilo, love.

Is that OK?'

Customer: 'Yes, that's fine, thanks.'

You see, while the customer is still allowed to use the p****
word, it is now against the law for the retailer to use imperial
units. If caught in this felonious act, the powers that be will
descend on him like a tonne of bricks. (That's the metric tonne,
of course, which is 98.3 per cent of the imperial ton.)

Angles

How to Name Angles (∠s)

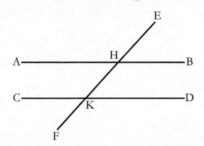

Often angles can be named by one letter only, but in the above diagram there are four angles round the point H. First, you write ∠H but then you have to say, 'Where is it coming from, where is it going to?' The top right angle is coming from E and going to B (or the other way round), so that angle is ∠EHB or ∠BHE. The angle below it is also at point H but it is coming from K and going to B, so it is ∠KHB or ∠BHK. The other angles at H are ∠KHA and ∠AHE.

Now name the angles round point K top right and going clockwise.

Answers: (1) ∠HKD (2) ∠FKD (3) ∠FKC (4) ∠HKC

Angles Made by Parallel Lines

Parallel (shorthand ∥) lines are lines that will never meet however far they are produced. A good visual aid for parallel lines is a railway track. When two parallel lines are cut by another line, this line is called a transversal.

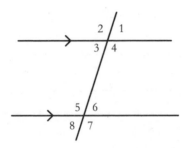

If one angle is known, then you can calculate all of them. Suppose ∠1 = 70°, then ∠2 = 110°, ∠3 = 70°, ∠4 = 110° and ∠s 5, 6, 7 and 8 are 110°, 70°, 110° and 70°, respectively. It's pretty obvious, isn't it?

But they all have fancy names:

(a) ∠s 1 and 3, for example, are called vertically opposite angles, and vertically opposite angles are equal. OK, they aren't actually 'vertical' – more diagonally opposite, perhaps, but please, don't quibble – just tilt your head to the side and try to see them as vertical.

(b) ∠s 1 and 2, for example, are called supplementary angles and they add to 180°.

(c) ∠s 3 and 6, and ∠s 4 and 5 are called alternate angles. They are equal and make a Z shape, backwards, forwards or upside down.

(d) ∠s 4 and 7, for example, are called corresponding angles. They are equal and make an F shape (again backwards, forwards or upside down).

(e) ∠s 3 and 5 and ∠s 4 and 6 are called co-interior angles (inside together) and they add to 180°.

All this basic, and rather boring, geometry – and most of the geometry to follow – was included in *Elements*, the thirteen profound tomes written by Euclid of Alexandria, who lived around 300 BC. Although not much is known about Euclid himself, Euclidean geometry is still studied in schools to this day but, mercifully, without the formal proofs, propositions and axioms that students were still being subjected to when I was at school in the mid-1950s.

At the end of every proof, we always wrote QED, which stood for '*quod erat demonstrandum*, the Latin for 'that which has been demonstrated'. Needless to say, none of us had a clue what it was all about.

There is a story about Euclid that has survived. Ptolemy Soter, one of the great Macedonian kings, ordered special roads to be built to facilitate quick and easy travel for couriers on the king's business. Ordinary people were not allowed to use these roads. Ptolemy Soter sponsored Euclid and was interested in his work but, like students twenty-four centuries later, the king found it unfathomable and so he asked Euclid to make it easier. 'Alas, Your Highness,' replied Euclid, 'I regret to tell you there is no royal road for geometry.'

Another Terrible Maths Joke

Parallel lines have so much in common. It's a shame they will never meet.

The Triangle

There are various ways to describe a triangle. Sometimes they are described by the size of their angles: right-angled triangles have one angle of exactly 90^0, acute-angled triangles have all three angles sharp and less than 90^0 and obtuse-angled triangles have one angle wide and greater than 90^0.

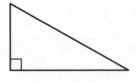

A right-angled triangle

An acute-angled triangle

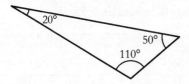

An obtuse-angled triangle

Triangles are also described by their sides: equilateral, with all sides and angles equal; isosceles, with two sides and two angles equal; and scalene, with all sides a different length.

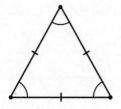

An equilateral triangle

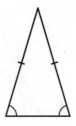

An isosceles triangle

To draw an isosceles triangle, draw the base the required length. With your compass stretched out to the length you wish the other two sides to be, put the point of the compass at the end of the base line and draw an arc above. Repeat this at the other end of the line so that the arcs cross. Join the lines.

For an equilateral triangle, draw the base line but this time, before you make the arcs, see that the distance between the compass point and the pencil point is the same length as the base.

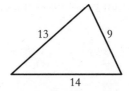

A scalene triangle
(again you can draw it with a ruler and compass)

An Imposter

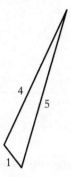

This skinny 'thing' is not a triangle at all. It is masquerading as a triangle and should be reported to the maths police. Look at it. The longest side of the 'thing' is 5 units and the other two sides are *marked* 4 and 1. But if they are really 4 and 1, they would collapse on top of the longer side exactly and there would be no triangle.

Likewise, there can be no triangle with sides of 3, 4, 8 or 2, 5, 2, or an infinite number of other impossible triples.

The Inequality Rule for All Triangles

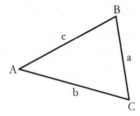

All triangles, whether they be obtuse-angled, acute-angled, right-angled, equilateral, isosceles or scalene, have to obey this rule:

$$a + b > c$$
and
$$a + c > b$$
and
$$b + c > a$$

Otherwise no triangle is possible.

Maths Joke – Even Worse Than Usual

Which triangle is the coldest? An ice-sosoles triangle.

The Sum of the Angles of a Triangle

The angles of every triangle add up to 180^0. To prove this, draw any triangle, large or small, and write A, B and C inside the corners. Cut it out and tear off the corners. Then lay them out in jigsaw-puzzle fashion and, hey presto, the angles lie in a straight line or 180^0.

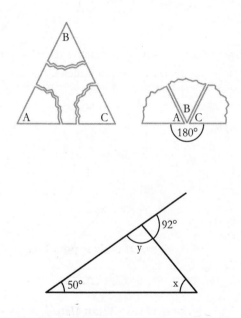

What is the size of ∠s *x* and *y*?
Answer: y = 88° and *x* = 42°

The Area of a Triangle

There are a few formulae for calculating the area of a triangle, depending on the data given.

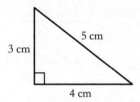

The little square in the corner reminds you that the triangle is right-angled. So the area of the triangle is:

½ base × height = ½ × 4 × 3 = 6 cm² (the 5 is not required).

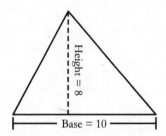

In this triangle, the area is ½ base × height again. So:

A = ½ × 10 × 8 = 40 units².

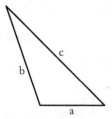

This triangle is scalene with a = 3 cm, b= 7 cm and c = 8 cm.

Finding the area is a bit more complicated. But we use the two-thousand-year-old formula discovered by Greek mathematician Heron of Alexandria:

$$A = \sqrt{s(s - a)(s - b)(s - c)}$$

Step 1: First calculate s, the semi-perimeter, which is:

$$\frac{1}{2}(a + b + c)$$
$$= \frac{1}{2}(3 + 7 + 8)$$
$$= \frac{1}{2} \times 18$$
$$= 9$$

Step 2: Now calculate s(s –a)(s – b)(s – c), which is:

$$9(9 - 3)(9 - 7)(9 - 8)$$
$$= 9 \times 6 \times 2 \times 1$$
$$= 108$$

Step 3: Calculate $\sqrt{108}$ = 10.4 (with your calculator, key in 108, $\sqrt{x}$).

So the area of the scalene triangle is 10.4 cm^2.

Triangular Numbers

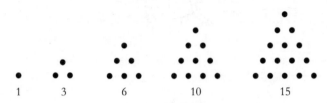

These triangular numbers are associated with the little genius who astounded his teacher by adding the numbers from 1 to 100 in a few seconds.

They go on forever but you never need to count the spots. Using the same formula Gauss used for his calculation, the next triangle along the line will be $6 \times 7 \div 2 = 21$ and the next one after that $7 \times 8 \div 2 = 28$, or, putting this in algebra, $n \times (n + 1) \div 2$.

Of course, 1 is only a dot, but it must be included as $1 \times 2 \div 2 = 1$. (However, 1 was excluded from the prestigious prime number club – see later.)

Notice that every two adjacent triangular numbers make a square number. For example, $6 + 10 = 16 = 4^2$, $10 + 15 = 25 = 5^2$, $15 + 21 = 36 = 6^2$, and so on, along the never-ending number line.

PLAYING THE TRIANGLE, THE SQUARE AND THE CIRCLE

The Square

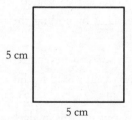

5 cm

5 cm

The area = 5^2 = 25 cm². Easy! I bet you knew that already. But what if the square is 100 m by 100 m? The area is 100 × 100 = 10 000 m² and that is 1 hectare.

Trafalgar Square in London is 1 hectare of attractive public space where tourists love to congregate.

Union Square in San Francisco is a little more than a hectare. Like Trafalgar Square, it is a wonderful place to meet up with friends, shop, dine and see a show.

The Rectangle

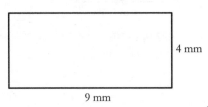

4 mm

9 mm

The area of the rectangle is $9 \times 4 = 36$ mm^2, and again you probably know that. However, if the rectangle has sides 22 yards by 220 yards, the area is 4840 yards^2, and 4840 square yards equals 1 acre.

But 22 yards equals 1 chain and 220 yards = 1 furlong, so a chain multiplied by a furlong equals 1 acre. In days gone by, an acre was considered to be the area of land that could be ploughed by a yoke of oxen in one day.

The gardens at Buckingham Palace measure 42 acres and that includes a helicopter-landing pad.

So the hectare is the metric unit for measuring land and the acre is the imperial unit. A hectare is approximately two and a half times an acre. Both are used in the UK.

The Parallelogram

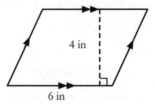

The parallelogram has opposite sides parallel and opposite angles equal. The area formula is base times perpendicular height, so the area of the above shape is 6 × 4 = 24 in².

The Rhombus

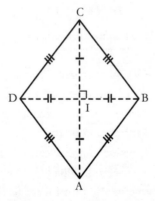

The rhombus, with all its sides equal and opposite sides parallel, is a squint square and the formula for its area is half of one diagonal times the other diagonal. If DB = 10 cm and AC = 14 cm, the area of the rhombus is ½ × 14 × 10 = 70 cm².

The Kite

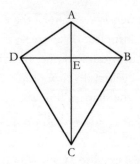

The formula for the area of a kite is exactly the same as for a rhombus.

The Trapezium

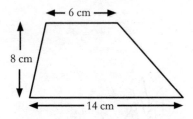

The trapezium (in the US, trapezoid) has two sides parallel and nothing else special about it. The area is half the sum of the parallel sides times the vertical distance between the parallel sides. So the area is $\frac{1}{2}(14 + 6) \times 8 = 10 \times 8 = 80$ cm^2.

Lastly

If you double the dimensions of any of these shapes, the area is not doubled. It is quadrupled. If you treble the dimensions, you get a similar shape but with nine times the area.

Putting this into algebra, if you multiply the sides or diagonals by x, the area of the shape will be x^2 as big.

An Appallingly Terrible Maths Joke

Question: What do you call a crushed angle?

Answer: A wreckedangle.

A WRECKED AN▸GLE

Pythagoras

Pythagoras is best known for his famous theorem. A theorem is a mathematical statement that has been proved beyond doubt by the mathematician himself or other mathematicians. A conjecture may also be a brilliant mathematical idea but it has yet to be proved.

Pythagoras was born on the Greek island of Samos and lived from about 580 to 500 BC. That's roughly 2560 years ago, so many of the stories about him may have grown arms and legs. His life and ideas are shrouded in myth and exaggeration. But since there is no fear of being sued for libel, it is fun for thirteen-year-olds who are studying the famous theorem to hear about his eccentricities.

He was the leader of a group of like-minded oddballs. Unusual for his days, there were women in the commune. Theano, the wife of Pythagoras, was also obsessed with numbers but, of course, she never had a theorem. They were all vegetarians but beans were strictly forbidden. Pythagoras considered that beans resembled tiny human foetuses and to eat them would be cannibalism.

He also believed that human beings are reincarnated every 216 years.

Instead of worshipping the Greek gods – Zeus, Poseidon,

Apollo, Aphrodite, Hermes, Hades et al. – they worshipped numbers. All even numbers were female and earthly, and (of course!) all odd numbers were male and divine.

Their favourite shape was the pentagram with the top pointing to the stars and the other four points representing earth, air, fire and water.

Their favourite solid was the sphere and for that reason only they believed, correctly, that the world was round.

No one knows how Pythagoras came up with his famous theorem. One suggestion is that when sitting on his patio in the beautiful Greek sunshine thinking deep philosophical thoughts, the marble tiles caught his eye. They were in the pattern below.

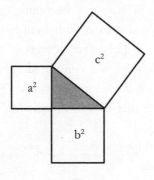

Perhaps he said to himself, 'Ah, ha! I wonder if the square on the longest side of that right-angled triangle equals the sum of the squares on the other two sides?' And it was! And he proved it! Except, he called the longest side 'hypoteinousa', which is Greek for 'hypotenuse'.

But before you try some Pythagoras calculations, you must learn about square roots.

Square Roots

The square root of any number n is the number that gives n when multiplied by itself. For example:

$$\sqrt{100} = 10 \text{ because } 10 \times 10 = 100$$
$$\sqrt{49} = 7 \text{ because } 7 \times 7 = 49$$
$$\sqrt{144} = 12 \text{ because } 12 \times 12 = 144$$

Harder square roots are easily done on your calculator. What is $\sqrt{5041}$? Key in 5041 and press the $\sqrt{x}$ button. Up pops 71. It's as simple as that.

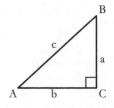

In any triangle ABC, the sides are normally named AB, BC and AC, but when doing Pythagoras calculations, it is usual to name them in lower-case letters, a, b and c opposite angles, A, B and C.

So $c^2 = a^2 + b^2$. What is c if a = 3 and b = 4?

$$c^2 = 3^2 + 4^2 = 9 + 16 = 25$$
$$c = \sqrt{25} = 5$$

So you have discovered for yourself the famous 3, 4, 5 Pythagorean triple. Even in the twenty-first century, carpenters and builders still use the 3, 4, 5 triangle to get accurate square corners.

What is c if a = 5 and b = 12?

$$c^2 = 5^2 + 12^2 = 25 + 144 = 169$$
$$c = \sqrt{169} = 13$$

The 5, 12, 13 is the second most famous triple.

Other famous triples you should verify and remember, are:

8, 15, 17
7, 24, 25
20, 21, 29
9, 40, 41

plus an infinite number of others.

And, of course, every triple can be 'scaled up'. So the basic 3, 4, 5 triple can become 6, 8, 10 or 30, 40, 50, and again an infinite number of others.

The Pythagoreans would have enjoyed discovering these new triangles. But there was trouble brewing.

A party of Pythagoreans were having a day out in a boat. A young student named Hippasus was having fun with a right-angled triangle with shorter sides of 1 and 1. Therefore the length of the hypotenuse was $\sqrt{2}$ and he couldn't get an exact answer. He got to 1.41, but 1.41 × 1.41 was only 1.9881. He suspected that $\sqrt{2}$ was a number that would never work out exactly and the decimal places would go on forever. He reported this to Pythagoras, who was horrified. As far as he and the others on board were concerned, all numbers were perfectly behaved and a number that did not work out exactly did not exist. How dare Hippasus suggest that some numbers were less than perfect – and 2 of all numbers, the first female number and one of their all-time favourites!

So they threw him into the water and Hippasus couldn't swim.

Later, Pythagoras claimed it was Poseidon, the Greek god of the sea, who drowned poor Hippasus. Poseidon could not condone such heretical mathematical beliefs. Hmm . . . a likely story!

Another Maths Joke

A wee bit funny this time - but remember, normal people do not laugh at maths jokes.

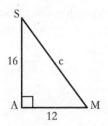

Exam question – find c

Student's Answer: look, it's there, falling down the slope.

Irrational Numbers

Hippasus had actually discovered irrational numbers. Irrational numbers cannot be expressed as a terminating decimal. For example, $\sqrt{2.25} = 1.5$ exactly, so $\sqrt{2.25}$ is a **rational** number, but $\sqrt{2} = 1.4142135\ldots$ and on and on, so $\sqrt{2}$ is an **irrational** number.

Likewise $\sqrt{6}$ is irrational but $\sqrt{6.25} = 2.5$ exactly and is therefore a rational number. Try them yourself with a calculator.

Do you notice how mathematicians pinch English words, in this case, rational and irrational, and give them an entirely different meaning?

The Death of Pythagoras

Pythagoras had a brutal death. A young, spoilt brat named Kylon, whose father was a rich nobleman, wanted to join the Pythagorean commune.

However, Pythagoras refused to admit him to the group because the youth was hopeless at mathematics. A humiliated Kylon assembled a mob to attack the Pythagoreans. Many were killed but Pythagoras broke free and ran for his life. A quick dash into a field might have saved him because he could have hidden in the foliage of the crop. But guess what the crop was! Yes, the field was planted with butter beans and Pythagoras could not bear to trample on even one of them.

So in 500 BC, Pythagoras was stabbed to death.

THE DEATH OF PYTHAGORAS

Cubes

For x^3 you say 'x cubed' but for any number above three, the little number is 'to the power of'. So 4^3 (4 cubed) means $4 \times 4 \times 4 = 64$; and 2^7 (2 to the power of 7) $= 2 \times 2 \times 2 \times 2 \times 2 \times 2 \times 2 = 128$; and 5^4 (5 to the power of 4) $= 5 \times 5 \times 5 \times 5 = 625$.

Here are the first ten numbers and their cubes. Try to memorise as many of them as possible.

1	2	3	4	5	6	7	8	9	10
1	8	27	64	125	216	343	512	729	1000

Note that every cube from 1 to 10 ends with a different digit and 1, 4, 5, 6, 9 and 10 have matching end digits.

An Easy Way to Add Cubes

This works for adding any number of consecutive cubes **starting at 1^3**. If you add the ten cubes above in the ordinary way, you get 3025.

However, an easier method is: $(1 + 2 + 3 + 4 + 5 + 6 + 7 + 8 + 9 + 10)^2$. Add 1 to 10 the Gauss way: $10 \times 11 \div 2 = 55$, and $55^2 = 3025$ ($5 \times 6 = 30$ and attach 25). Much easier, eh!

Cube Roots

The cube root of x is written $\sqrt[3]{x}$ so $\sqrt[3]{1000} = 10$. (Say to yourself, 'What times what times what gives 1000?')

Cube roots are not on most pocket calculators but there are many online calculators with a multitude of functions. For $\sqrt[3]{729}$, press 729 and the $\sqrt[3]{}$ button and there it is: 9.

By the way, remember Pythagoras believed that human beings are reincarnated every 216 years? Well, take his famous triple (3, 4, 5) and cube each number and add: $3^3 + 4^3 + 5^3 = 27 + 64 + 125 = 216$. Coincidence or what?

Four More Coincidences

$153 = 1^3 + 5^3 + 3^3$. This also works for 370, 371 and 407 but not for any other three-digit numbers.

x^0

Any number to the power of 0 equals 1, for example $2^0 = 1$, $7^0 = 1$, $100^0 = 1$, all except 0^0 and nobody in the whole wide world knows the answer to that! So we say 0^0 is undefined.

Surds

Surds? They sound nasty! But, as usual, the title of the topic sounds worse than the actual maths. You have met surds before. $\sqrt{2}$ is a surd and that was the number that led to the drowning of poor Hippasus.

Surds are numbers left in $\sqrt{x}$ form (or $\sqrt[3]{}$ or $\sqrt[4]{}$ etc.). For example, $\sqrt{5}$, $\sqrt{13}$, $\sqrt{18}$ are surds and, of course, they go on forever. But $\sqrt{9}$, $\sqrt{2.25}$, $\sqrt{121}$ are not surds because they are 3, 1.5 and 11 exactly.

Simplifying Surds

(a) $\sqrt{18} = \sqrt{9} \times \sqrt{2}$ and $\sqrt{9} = 3$. So $\sqrt{18} = 3\sqrt{2}$
(b) $\sqrt{50} = \sqrt{25} \times \sqrt{2}$ and $\sqrt{25} = 5$. So $\sqrt{50} = 5\sqrt{2}$

Try simplifying $\sqrt{8}$, $\sqrt{75}$ and $\sqrt{98}$.
Answers: $2\sqrt{2}$, $5\sqrt{3}$ and $7\sqrt{2}$

Rationalising a Denominator

Now, this really does sound difficult. But it is just a ploy you can use to make life easier. Suppose you had to calculate $\frac{1}{\sqrt{2}}$.

That would be $1 \div 1.4142 \ldots$ and the more decimal places, the more difficult it becomes. Before the invention of pocket calculators, it would have been mathematical torture. But help is at hand. We are going to change the irrational denominator to a simple rational number and this is how it is done.

You know that multiplying a number by 1 will not change the number; for example, 4×1 is still 4. Now $6 \div 6 = 1$. So $4 \times \frac{6}{6}$ is still 4. So we are going to multiply $\frac{1}{\sqrt{2}}$ by $\frac{\sqrt{2}}{\sqrt{2}}$. Multiplying the top line gives $1 \times \sqrt{2} = \sqrt{2}$ and multiplying the bottom line gives $\sqrt{2} \times \sqrt{2}$, which is $\sqrt{4} = 2$.

We haven't changed the value, but we have made a world of difference to the difficulty of the calculation. Now we have $\frac{\sqrt{2}}{2}$ which is

$$1.4142 \ldots \div 2 = 0.7071 \ldots$$

Let's try another one. What about $\frac{3}{\sqrt{5}}$? Multiply by $\frac{\sqrt{5}}{\sqrt{5}}$. The top

76

line is $3\sqrt{5}$ and the bottom line is $\sqrt{5} \times \sqrt{5} = \sqrt{25} = 5$, so we now have $3\sqrt{5} \div 5$ which is much easier.

Rationalise the following, leaving your answers in surd form:

(a) $\frac{2}{\sqrt{3}}$

(b) $\frac{5}{\sqrt{7}}$

Answers: (a) $\frac{2\sqrt{3}}{3}$ (b) $\frac{5\sqrt{7}}{7}$

The Golden Ratio

We find the Golden Ratio when we divide a line in two parts so that the longer part (AC) divided by the shorter part (CB) equals the whole length (AB) divided by the longer part (AC).

This line is not drawn to scale but if AC = 8.9 cm and CB = 5.5 cm, then

AC ÷ CB = 8.9 ÷ 5.5 = 1.6181 (to 4 decimal places, or 4dp)
AB ÷ AC = 14.4 ÷ 8.9 = 1.6179 (4dp)

The results are nearly equal and the above line is a good approximation to the Golden Ratio.

Opposite is a golden rectangle which has been divided to make a square and another smaller, but equally perfect, golden rectangle.

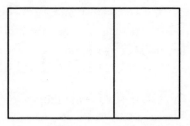

Measure the length and breadth of the two rectangles and divide the longer side by the shorter side in each case.

Both results should be 1.6(ish). Mathematicians call this number 'phi', using its symbol φ, the twenty-first letter in the Greek alphabet. However, your result will not be exact because φ is another of these irrational numbers where the decimal places go on and on forever.

$$φ = 1.618033988749894. . . \text{ forever.}$$

Another way of writing phi is $\frac{1 + \sqrt{5}}{2}$. Calculate $\sqrt{5}$ to as many decimal places as your calculator will allow. Add 1 and divide by 2 and the result will be the same as the above irrational number. Also $φ - 1 = \frac{1}{φ}$ (confirm it with your calculator).

Artists, architects, designers and photographers believe these golden rectangles are the most pleasing and beautiful shape possible and mathematicians have been studying them for thousands of years.

And then there is the wonderful pentagram. The Pythagoreans didn't only worship the pentagram, they also studied it and discovered that it held the Golden Ratio.

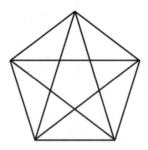

This pentagram within a pentagon is chock-a-block with lines in the Golden Ratio. Do some measuring and find the ratio for yourself. Look also for parallel lines. Measure them as well and do a few calculations.

The Golden Ratio continues to appear in modern everyday life. The length and breadth of bank cards, library cards, store cards and many others are approximately 86 mm by 54 mm, and 86 ÷ 54 is nearly φ. It is no coincidence that the slots in your purse or wallet are designed to hold these cards exactly.

The dimensions of many modern logos, such as the yellow rectangular border on the *National Geographic* cover and the striking oval of the Toyota trademark, are in the Golden Ratio.

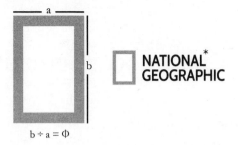

$b \div a = \Phi$

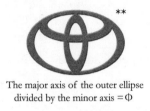

The major axis of the outer ellipse
divided by the minor axis $= \Phi$

Look around and see what else might be in the Golden Ratio.
Try books, postcards, playing cards, your widescreen TV –
perhaps the Renoir on your dining-room wall?

*National Geograpic and the associated logo are registered trade marks of
National Geographic Society.

**This logo is a registered trade mark of Toyota Motor Corporation.

The Fibonacci Sequence

Leonardo Fibonacci of Pisa, who lived from 1170 to 1250, was the most talented mathematician of his time. He was mentioned briefly at the beginning of this book. In *Liber Abaci* (Book of Calculations) he introduced Europe to 0–9, the digits we use today. Before that, everyone in Europe struggled with Roman numerals.

But he is best known for the Fibonacci Sequence. The pattern starts as follows:

1 1 2 3 5 8 13 21 34 55 89 144 233 377 610 987 1597 2584 4181. . .

and it goes on and on forever. (Each number is the sum of the two before.)

There are many interesting facts about the famous sequence. For example, notice 144. It is the twelfth number along and it has been proved that it is the only square in the whole series.

Add any ten consecutive numbers and you will find that the total is always divisible by 11.

But here is something more important and astonishing. Avoiding the first few numbers, take any number and divide it by the one before, for example, 1597 ÷ 987 and there it is again, the Golden Ratio, φ = 1.618 (3dp).

Mathematicians find many uses for the Fibonacci Sequence, but a practical one for us is a quick way to change between miles and kilometres. For example,

$$34 \text{ km} \approx 21 \text{ miles}$$
$$55 \text{ miles} \approx 89 \text{ km}$$

So, in fact, km: miles is approximately in the Golden Ratio.

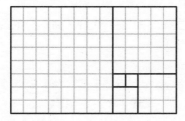

Above is Fibonacci's Golden Rectangle with squares whose sides are 1, 1, 2, 3, 5 and 8, but it can be extended to 13, 21 . . . on a bigger sheet of squared paper.

Fibonacci and Pythagoras

There is a weird connection between sixth-century BC Pythagoras and thirteen-century AD Fibonacci. Let's take any section of the Fibonacci Sequence again:

$$\ldots 13 \quad 21 \quad 34 \quad 55 \quad 89 \quad 144 \quad 233 \quad 377 \ldots$$

Select any group of four numbers, for example 21, 34, 55, 89. Multiply the outside numbers, $21 \times 89 = 1869$, and then multiply the two inside numbers and double it, $34 \times 55 \times 2 = 3740$, and miraculously they are the shorter sides of a right-angled triangle. Using the Theorem of Pythagoras (you will definitely need your calculator!):

$$c^2 = a^2 + b^2$$
$$= 1869^2 + 3740^2$$
$$= 3\,493\,161 + 13\,987\,600$$
$$= 17\,480\,761$$
$$c = \sqrt{17\,480\,761}$$
$$= 4181$$

So, to add to your collection of Pythagorean triples, you have (1869, 3740, 4181). And 4181 is the nineteenth Fibonacci number!

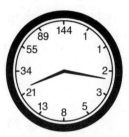

Sixteen minutes past eight
on a Fibonacci clock.

A Fibonacci-style Number Trick

Ask a friend to write any two numbers and then 'do a Fibonacci' and finish with ten numbers altogether. For example, they might start with 2 and 8, getting: 2, 8, 10, 18, 28, 46, 74, 120, 194, 314.

Ask them to add the ten numbers but not to tell you the total. All you need to know is their seventh number. In this case it is 74. Multiply 74 by 11 and you have 814, exactly the total when the ten numbers are added.

You friend may have started with 8, then 2. This time there will be a different sequence: 8, 2, 10, 12, 22, 34, 56, 90, 146, 236. This time, the total is 616, but you will take the seventh number again and multiply it by 11, getting 56 × 11 = 616.

If you would like to see why this works, 'do a Fibonacci' with chosen numbers a and b and add them:

$$a$$
$$b$$
$$a + b$$
$$a + 2b$$
$$2a + 3b$$
$$3a + 5b$$
$$\mathbf{5a + 8b}$$
$$8a + 13b$$
$$13a + 21b$$
$$\underline{21a + 34b}$$
$$\mathbf{55a + 88b}$$

Take the seventh number, **5a + 8b**, multiply it by 11, and you get the total **55a + 88b**. There is really no trickery in maths – only magic.

Fractions

Fractions are not easy and yet they crop up in everyday conversation more than any other topic in maths. They are made even more difficult by the language that comes with them. A long time ago, some unknown mathematician with a wicked sense of humour invented words and phrases like numerator, denominator, cancelling, proper and improper, mixed numbers, factors, highest common factor (HCF) and lowest common multiple (LCM).

However, the culprit with the love of long mathematical words might not have been British. In an old French maths textbook, I read the helpful information, *Dans une fraction, le numérateur se trouve au-dessus du dénominateur.* The translation is roughly, 'In a fraction, the numerator is above the denominator.' And the French also have *les fractions impropres.*

So having guessed that an improper fraction is probably not downright indecent, what is it? It is simply a top-heavy fraction like ⁵⁄₂ or ⁷⁄₃ where the numerator is bigger than the denominator.

Another difficulty! The addition and subtraction of fractions is done in a completely different way from multiplying and dividing fractions. You can't do fractions on a simple pocket calculator, but if all else fails google 'fraction calculators' and you can choose from dozens of them.

That, though, is defeatist talk. Come on, give it a try!

Adding Fractions

Let's get the maths jargon sorted first. For adding and subtracting fractions, you need to know:

(a) Mixed number. That is a combination of a whole number and a fraction like $4\frac{2}{3}$ or $5\frac{1}{4}$. You often need to change mixed numbers to improper fractions and vice versa. For $4\frac{2}{3}$, you say '3 times 4 = 12 and add 2' and you get the improper fraction $\frac{14}{3}$. For $5\frac{1}{4}$, you say '4 times 5 = 20 and add 1', getting $\frac{21}{4}$. To change an improper fraction, e.g. $\frac{23}{7}$, to a mixed number, you say '7 into 23 is 3 and 2 over', so the mixed number is $3\frac{2}{7}$.

(b) You also need to know about LCM: the lowest common multiple. What is the LCM of 8 and 6? Method: say the times table of the larger number until the other number divides in exactly. So, 8, 16, 24 and stop there because 6 goes into 24. The LCM of 6 and 8 is therefore 24. One more: what is the LCM of 12, 15 and 20? Say the 20 times table until both the other numbers divide in. OK, 20, 40, 60 and stop there because both 12 and 15 go into 60. So the LCM of 12, 15 and 20 is 60.

Now you are ready to add fractions. Examples:

(a) $\frac{2}{3} + \frac{1}{4}$

Find the LCM of 3 and 4

$$= \frac{8}{12} + \frac{3}{12}$$
$$= \frac{11}{12}$$

(b) $\frac{4}{5} + \frac{3}{4}$

$$= \frac{16}{20} + \frac{15}{20}$$
$$= \frac{31}{20}$$
$$= 1\frac{11}{20}$$

(c) $9\frac{2}{5} + 7\frac{1}{3}$

Add the whole numbers first

$$= 16 + \frac{6}{15} + \frac{5}{15}$$
$$= 16\frac{11}{15}$$

(d) $3\frac{9}{10} + 2\frac{3}{4}$

$$= 5 + \frac{18}{20} + \frac{15}{20}$$
$$= 5 + \frac{33}{20}$$

But $\frac{33}{20}$ is improper, so change it to a mixed number = $1\frac{13}{20}$, making your answer $6\frac{13}{20}$.

Congratulations! You have mastered the adding of fractions.

A Truly Terrible Fractions Joke

Question: Who invented fractions? Answer: King Henry ⅛.

HENRY THE $\frac{1}{8}$

Subtracting Fractions

Subtracting fractions is much the same as adding them. Remember to subtract the whole numbers first:

$$\text{(a) } 4\tfrac{2}{5} - 2\tfrac{1}{10}$$
$$= 2 + (\tfrac{2}{5} - \tfrac{1}{10})$$
$$= 2 + (\tfrac{4}{10} - \tfrac{1}{10})$$
$$= 2\tfrac{3}{10}$$

(b) $5\frac{1}{2} - 1\frac{7}{8}$

$= 4 + (\frac{1}{2} - \frac{7}{8})$

$= 4 + (\frac{4}{8} - \frac{7}{8})$

Ah-ha! You can't take 7 from 4.

So steal 1 from 4, call it $\frac{8}{8}$ and give it to $\frac{4}{8}$, making $\frac{12}{8}$.

So now you have $3 + (\frac{12}{8} - \frac{7}{8})$

$= 3\frac{5}{8}$

Try $3\frac{1}{3} - 1\frac{1}{2}$ and if you get $1\frac{5}{6}$, you have cracked it!

Multiplying Fractions

Again we will get the maths jargon sorted first.

Let's start with **factor** and **highest common factor** (HCF). In maths, factor just means divisor. So the factors of 12 are 1, 2, 3, 4, 6, 12 and the factors of 18 are 1, 2, 3, 6, 9, 18, so the HCF, the highest divisor *common* to both 12 and 18, is 6.

Cancelling in normal English means deciding that a planned event will not take place, like a wedding or a pop concert. In maths, it means simplifying a fraction. For example, if a cake is cut into six slices and you eat three of them, then you have devoured $\frac{3}{6}$ or a half. So you use the factors to cancel or, in layman's terms, use divisors to make the fraction simpler.

But suppose you are faced with $\frac{63}{84}$, it doesn't matter if you don't spot that the HCF was 21. Divide the numerator and denominator by 3 first, getting $^{21}\!/_{28}$, and then you will see that 7 is a divisor also, getting $\frac{3}{4}$.

Now, multiplying and dividing fractions are **not** done the same way as adding and subtracting. You do **not** multiply the whole numbers first. Mixed numbers must first be changed to improper fractions (see p. 88). Then it gets easy. Examples:

(a) $\frac{1}{3} \times \frac{1}{4}$

Multiply the top line and multiply the bottom line and you get $^{1}\!/_{12}$.

It's as easy as that.

(b) $\frac{2}{5} \times \frac{3}{10}$

$= ^{6}\!/_{50}$

$= ^{3}\!/_{25}$

$2\frac{2}{3} \times 1\frac{1}{8}$

$= \frac{8}{3} \times \frac{9}{8}$

Cancel by 8 top and bottom,

getting $^{9}\!/_{3} = 3$

Try $3\frac{1}{3} \times 1\frac{1}{5}$ and if you get 4 exactly, you have passed the test.

A Few Lines of Poetry

Mathematicians like poetry too, you know. Here are a few lines from Alfred, Lord Tennyson's famous poem 'The Charge of the

Light Brigade' after the ill-fated Battle of Balaclava during the Crimean War:

> Theirs not to make reply,
> Theirs not to reason why,
> Theirs but to do and die.
> Into the Valley of Death
> Rode the six hundred.

The maths class version of the poem is:

> Ours not to make reply,
> Ours not to reason why,
> Dividing fractions? Don't even try!
> Flip them round and multiply.

Ah well, maybe we are lowbrow philistines but the little poem will help you remember how to tackle dividing fractions.

Dividing Fractions

The above poetry, both the classical stuff and the little ditty, suggests that we should just do as we are told and not ask questions. But you might be asked sometime, 'Why do you turn the second fraction upside down?' and as a born-again mathematician, you need to know.

It's simple! If you have a pizza and you divide it into quarters, how many people will get a slice? Four, of course.

$$1 \div \tfrac{1}{4} = 1 \times \tfrac{4}{1} = 4$$

I bet you are fond of those delicious bars of Swiss milk chocolate with the honey and caramel nougat. You buy four bars but decide to eat only ⅖ of a bar each day. If you have the will-power to ration yourself to that, how many days will the chocolate last? Well, that's $4 \div \tfrac{2}{5} = 4 \times \tfrac{5}{2} = \tfrac{20}{2} = 10$ days. So it makes sense, doesn't it? Do you notice, and this is important, that when you divide by a fraction, the answer is always bigger?

Let's try one more with mixed numbers:

$$4\tfrac{1}{8} \div 1\tfrac{1}{10}$$
$$= {}^{33}\!/_{8} \div {}^{11}\!/_{10}$$
$$= {}^{33}\!/_{8} \times {}^{10}\!/_{11}$$
$$= {}^{330}\!/_{88}$$

(Cancel by 22 or 11 and then 2)
$${}^{15}\!/_{4}$$
$$= 3\tfrac{3}{4}$$

Understanding $x \div 0$

At the beginning of the book you learned that $x \div 0$ was impossible to calculate. Having mastered dividing by fractions, you will now understand why this is so. Let x equal any number, say 5. Now, using our little ditty, 'Divide by $\frac{1}{10}$? Don't even try! Flip it round and multiply':

$$5 \div \tfrac{1}{10} = 50$$
$$5 \div \tfrac{1}{100} = 500$$
$$5 \div \tfrac{1}{1000} = 5000$$
$$5 \div \tfrac{1}{1\,000\,000} = 5\,000\,000$$

You see that the smaller the fraction, the larger the answer. So when the fraction is the smallest number you can think of, the answer will be the biggest number you can think of. But nobody can think of the very lowest number because there isn't one, nor is there such a thing as the largest number. So that is why there is no answer for $x \div 0$ and mathematicians say it is undefined.

Negative Powers

You probably have no idea what 2^{-3} means. But it's easy. When you see a negative index number, say '1 over' and remove the – sign. So 2^{-3} = 1 over 2^3 = 1 over 8, which is the fraction ⅛. Likewise, 3^{-2} is 1 over 3^2, which is the fraction ⅑.

Try these:

> (a) 5^{-2}
> (b) 4^{-1}
> (c) 6^{-3}

Answers: (a) ¹⁄₂₅ (b) ¼ (c) ¹⁄₂₁₆

Factorial!

If you own a scientific calculator or can access one of many online, there will be a button you may not have noticed before. It is $x!$ In written English, there is quite a difference between 'Harry won', which is merely a statement, and 'Harry won!', which suggests 'Wow! Did he really? Well, good for Harry!'

So, in maths, surely 6! means 'Gosh! As many as six? I never knew he scored all six!'

Sorry to disappoint you. In maths, the ! symbol means 'factorial', so 6! (pronounced 6 factorial) means $6 \times 5 \times 4 \times 3 \times 2 \times 1$. So 6! = 720. 1! = 1; 2! = 2×1 = 2; 3! = $3 \times 2 \times 1$ = 6; 8! = $8 \times 7 \times 6 \times 5 \times 4 \times 3 \times 2 \times 1$ = 40 320; and by the time you get to 10!, the answer is over three million. It's simply shorthand for writing a string of decreasing whole numbers that need to be multiplied. All factorial numbers are even except 1!

Dividing factorials is so easy. Example: 100! ÷ 98! would be impossible on any pocket calculator but you can do it in seconds:

$$100! \div 98!$$
$$= \frac{100!}{98!}$$
$$= \frac{(100 \times 99) \times 98!}{98!}$$

Cancel 98! top and bottom
$$= 100 \times 99$$
$$= 9900$$

The National Lottery

From October 2015, the UK National Lottery decided to 'create more chances to become millionaires than ever before'. We must now choose our six lucky numbers from 59, instead of 49 numbers. Wow! Surely that was good news, wasn't it? More choice is always a good thing, don't you think? And we still have to pay only £2 for each ticket! Before the above date, our chance of becoming a multimillionaire was approximately 1 in 14 000 000. What is our chance now? The calculation is done with factorials.

$$59! \div (53! \times 6!)$$
$$\frac{59 \times 58 \times 57 \times 56 \times 55 \times 54 \times 53!}{53! \times 6!}$$

Cancel 53! top and bottom, getting
$$\frac{59 \times 58 \times 57 \times 56 \times 55 \times 54}{6 \times 5 \times 4 \times 3 \times 2 \times 1}$$

6 will cancel with 54, 5 with 55, 4 with 56, 3 with 57 and 2 with 58.
So you are left with $59 \times 29 \times 19 \times 14 \times 11 \times 9$
$$= \mathbf{45\,057\,474}$$

Our chance of winning the jackpot is more than three times **worse** than it was before. Don't expect to give up your day job any time soon.

Something You Really Need to Know

10! seconds equals six weeks exactly.

(This statement really needs an exclamation mark, but to avoid confusion it has been omitted.)

Proof: 6 weeks = 6 × 7 days = 6 × 7 × 24 hours = 6 × 7 × 24 × 60 minutes = 6 × 7 × 24 × 60 × 60 seconds = 3 628 800 seconds. 10! = 3 628 800 (on your calculator, press 10, $x!$).

If you are still at school, knowing this fact will make your six weeks' summer holiday seem even longer.

Shuffling Cards

Would it be possible to shuffle a pack of cards and get the cards in perfect order, Hearts, Clubs, Diamonds, Spades, with each suit from Ace to King? Well, it would be the biggest miracle ever because the chance is 1 in 52! (No, no, not 1 in 52 exclamation mark; it's 1 in 52 factorial!) That is $52 \times 51 \times 50 \times 49 \times \ldots \ldots \ldots 3 \times 2 \times 1$, which is an enormous number with over 70 digits.

A Clever Little Factorial to Finish With

$$1! + 4! + 5! = 145$$
(Test it for yourself.)

A Little More Pi

Even if you did have a go at finding pi after reading the introduction to this book, try once more. Perhaps you have a circular table or a circular lampshade or a circular biscuit tin?

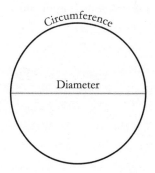

Circumference

Diameter

Measure round the circumference and across the diameter with a tape measure and calculate C ÷ D as before and hopefully you have got 3.1(ish).

There is a mention of the special number in the Old Testament of the Bible (1 Kings 7:23): 'And he made a molten sea, ten cubits from one brim to the other: it was round all about, and its height was five cubits: and a line of thirty cubits did compass it round about.' The cubit was the length of a man's forearm from the elbow to the tip of the middle finger and varied from about 1 foot 6 inches to 1 foot 9 inches.

So it couldn't have been an actual sea the author was writing about. In fact, it was a huge, round, ceremonial basin made of molten copper by order of King Solomon to be used for the ablution of his priests. If they had been interested in the maths and divided 30 cubits by 10 cubits, the result would have been 3 exactly.

The Egyptians and the Babylonians knew about this number and they thought it was 3 and a bit.

Archimedes of Syracuse lived in the third century BC and is still considered to be the world's greatest mathematician, scientist and engineer. He had a go at calculating this special number. He knew he hadn't got an exact result but he was sure it was somewhere between $^{223}/_{71}$ and $^{22}/_{7}$ – in both cases, 3.14 (2dp). Using this value, however, he also discovered and proved the formulae for finding the volume and surface area of a sphere.

Throughout the ages, mathematicians have been fascinated by this number but it wasn't until the early eighteenth century that William Jones, a little-known Welsh mathematician, introduced the Greek letter π (pi), the sixteenth letter in the Greek alphabet. Slowly, but surely, it took the place of the clumsy reference (usually in Latin): 'the quantity which, when the diameter is multiplied by it, yields the circumference.'

Jones also suspected that π was irrational but he couldn't prove it. That all-important proof did not come until the middle of the nineteenth century, but when it came, the flood-gates opened! It was a new obsession for mathematicians. How many digits of pi could they calculate?

At that time, William Shanks, a school headmaster and amateur mathematician, calculated pi to 707 decimal places and discovered that the digit 7 did not occur as often as the other digits. This had mathematicians very puzzled and concerned as they had thought that the digits went on and on in a random fashion, with no obvious patterns and no digit appearing more or less than any other.

However, long after Mr Shanks went to that peaceful, pi-ful paradise in the sky, it was discovered that he had made a mistake at the 527th digit and that everything was beautifully random after all.

Since the advent of powerful computers, those pi obsessives have gone totally crazy. In October 2011, a Japanese IT specialist calculated 12 trillion digits. That's 12 000 000 000 000! And you will be interested to know that the ten-trillionth digit is 5.

Another mathematical sport is to memorise the digits of pi. But let's not go there. For most of us, memorising our mobile phone number is enough.

Back in real life, $\pi = 3.14$ is fine for everyday calculations, although scientists use several more decimal places for accuracy in their work. If you are working with fractions, $^{22}/_{7}$ is also useful.

If you really would like to memorise a few more decimal places, a good mnemonic is 'Wow! I made a great discovery'. Count the letters in each word, getting 3.14159.

Pi is popular. It's the pop star of mathematics. Pi is everywhere in our daily lives – in your can of beans, your wine glass, your wedding ring, and as you look around while reading this book, in numerous other places.

Pi lovers like to celebrate Pi Day every year. In the US, they write the date with the month first, then the day and year. So in 2015, 14th March was 3 14 15, the first five digits of π. So that was a really special Pi Day when, of course, they sang their special song, *Pi, Pi, Mathematical Pi*, to the tune of that great song everyone danced to in the 1970s, 'Bye-bye, Miss American Pie' by Don McLean.

Another of These Terrible Maths Jokes

3.14 per cent of sailors are Pi-rates.

Calculating the Area and Perimeter of Shapes Using π

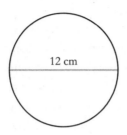

12 cm

The formula for the area of a circle is $A = \pi r^2$. In the above circle, d = 12 so r = 6.

$$A = 3.14 \times 6 \times 6$$
$$= 113 \text{ cm}^2$$

and the perimeter (or circumference) is

$$C = \pi d$$
$$= 3.14 \times 12$$
$$= 37.7 \text{ cm}$$

Another Appalling Maths Joke

Question: Which of King Alfred's knights invented the round table?

Answer: Sir Cumference!

The Theorem That Pythagoras Probably Never Knew

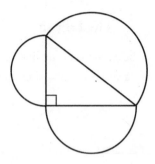

The area of the semi-circle on the hypotenuse of a right-angled triangle equals the sum of the areas of the semi-circles on the other two sides. Yes, it really does! Let's test it first with Pythagoras's best-known triple, 3, 4, 5.

The formula for the area of a semi-circle is $A = \frac{1}{2}\pi r^2$ where $r = 1\frac{1}{2}$, 2, and $2\frac{1}{2}$ respectively. So, if the Theorem of Pythagoras works for semi-circles, then:

$$\frac{1}{2}\pi(1\frac{1}{2})^2 + \frac{1}{2}\pi(2^2) = \frac{1}{2}\pi(2\frac{1}{2})^2$$

And it does, because $(1\frac{1}{2})^2 + 2^2 = 2.25 + 4 = 6.25$ and $(2\frac{1}{2})^2 = 6.25$.

Taking any other Pythagorean triple, for example (9, 40, 41), the areas of the semi-circles are:

$$\frac{1}{2}\pi(4.5)^2, \frac{1}{2}\pi(20)^2 \text{ and } \frac{1}{2}\pi(20.5)^2$$
$$4.5^2 + 20^2 = 20.25 + 400 = 420.25$$
$$\text{and } 20.5^2 = 420.25$$

The Ellipse

The area formula of the ellipse is rather like that of the circle. It is simply A = π ab where a and b are the major and minor semi-axes.

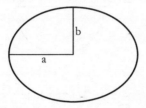

So if a = 10 cm and b = 7 cm, the area of the ellipse = 3.14 × 10 × 7 = 219.8 cm².

But, although the calculation of the area was easy, calculating the length of the perimeter is fiendishly difficult and could involve using this formula:

$$p = \pi\,(a + b) \sum_{n=0}^{\infty} \binom{0.5}{n}^2 h^n$$

where:

$$h = \frac{(a - b)^2}{(a + b)^2}$$

Eek! This scary, mysterious-looking formula gives the exact answer but it certainly will not be *Easy as Pi*, the promise made in the title of this book. It will make sense to only 1 in 20 000 people. So walk on by. Ignore it. Pretend you have never seen it.

Here is a Plan B.

The world-renowned Indian mathematician Srinivasa Ramanujan was born in 1887 and lived only thirty-two years, but in that short time he made an extraordinary contribution to mathematical analysis and number theory. He came up with three formulae for calculating the approximate length of the perimeter of an ellipse. His first and simplest one was:

$$ p \approx 2\pi \sqrt{\frac{a^2 + b^2}{2}} $$

where p is the perimeter and a and b, the semi-axes, are 10 cm and 7 cm as before.

$$ a^2 + b^2 $$
$$ = 10^2 + 7^2 $$
$$ = 149 $$

Now, divide by 2 and calculate the square root:

$$ \sqrt{74.5} $$
$$ = 8.63 $$

and multiply by 2π

$$= 8.63 \times 2 \times 3.14$$
$$= 54.2 \text{ cm (1 dp)}$$

Here is a challenge for you. Find an oval pie dish and measure the axes. Divide by 2 to find the semi-axes. Do the above calculation using your data to find the perimeter. Now, using your tape, measure the distance all round the perimeter of the pie dish. How does your measurement compare with your calculation?

These Eccentric Mathematicians

Ramanujan was typical of so many mathematicians. He lived for maths; nothing else was important. Lying terminally ill in an English hospital, he was visited by his friend, sponsor and fellow mathematician G. H. Hardy. Hardy remarked that he had travelled to the hospital by a taxi numbered 1729, in his opinion not a very interesting number. Ramanujan opened his eyes, raised his head from the pillow and pointed out to Hardy that 1729 was a very interesting number. It was the smallest number to be the sum of two cubes in two different ways, $1^3 + 12^3$ and $9^3 + 10^3$.

Carl Friedrich Gauss was so obsessed by numbers that he counted prime numbers for relaxation in the evening when he was finished with the 'heavy' maths at the university.

On one occasion, he was interrupted in the middle of a tricky mathematical problem by his wife's doctor and was informed that she was dying. He replied, 'Tell my dear wife to hang on for a little while longer until I finish this calculation.'

Archimedes, the father of mathematics, did not behave like a normal person when his bath overflowed. He didn't call for his slave to wipe up the mess. Instead he ran through the streets of Syracuse stark naked, shouting 'Eureka! Eureka!' He had discovered that any object, wholly or partially immersed in a fluid, is buoyed up by a force equal to the weight of the fluid displaced by the object. This is the Principle of Archimedes, chanted in every school physics laboratory in the world.

Years later, Syracuse was invaded by the Romans. Because of the ingenious weapons of war that Archimedes had invented, the siege had taken longer than Marcellus, the Roman general, had expected. So Marcellus gave orders that this brilliant man should not be killed but should be brought to him instead. However, Archimedes was drawing circles in the sand and deep in a mathematical problem. He refused to be disturbed. '*Noli turbare circulos meos*,' he said, 'Do not disturb my circles,' at which an angry soldier drew his sword and killed him.

And last, one of the most famous anecdotes in the history of mathematics. When Isaac Newton was enjoying a walk in his orchard, he was hit on the head by a falling apple. He didn't wince, rub his scalp and then mull over whether the apple was ripe enough to eat. Instead, he pondered about the universal laws of gravity.

In the seventeenth century, Newton and Gottfried Leibniz, (who also invented the binary numbers) independently invented calculus, the most difficult branch of mathematics taught in schools.

Infinity ∞

In the mind-boggling formula a little earlier, which I advised you to ignore, you will have seen the sign for infinity. ∞ is at the end of forever. It is not an actual number because there is no number at the end of forever. It is the concept of something so large that it is limitless and beyond our reach and imagination.

My love for you is infinite.

Sorry. I'm afraid I can't imagine that.

Tangents and Tangos

The word 'tangent' comes from the Latin *tangere*, which means 'to touch'. A few other English words come from the same derivation, for example, tangible, tactile and, in particular, that naughty dance, the tango.

Naughty? Yes, well, it was in 1910! You see, in the eighteenth and nineteenth centuries, dancing was very prim and proper. In the quadrille, four couples danced in a square. They sauntered around, changing partners, bowing to each other but never touching. In the Viennese waltz they danced ten inches apart, touching with gloved hands only.

The Argentine tango lives up to its Latin derivation. The dance came from the backstreets of Buenos Aires to Paris and then to London at the beginning of the twentieth century. It was a passionate dance in which the couples swayed, and hugged, their bodies and faces touching. To the narrow minded it was the epitome of decadence, but it was instantly popular and remains so to this day.

But there is nothing sexy about touching tangents:

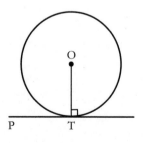

A tangent is a line that touches the circumference of a circle at one point only and is at right angles to the radius at the point of contact.

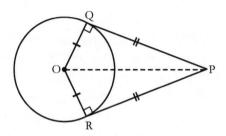

Example: If the length of the tangent is 6 cm and the radius is 2.5 cm, what is the length of OP? This is a Pythagorean triangle, so $OP^2 = \sqrt{(6^2 + 2.5^2)} = \sqrt{(36 + 6.25)} = \sqrt{42.25} = 6.5$ cm.

Angles in a Semi-circle

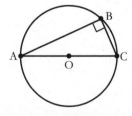

Any angle in a semi-circle is a right angle. This mathematical fact was discovered by Thales of Miletus, a mathematician and philosopher in Asia Minor (now Turkey). He lived from about

624 to 545 BC. It is difficult to imagine how he even managed to draw an accurate diagram all these years ago. Now, in the twenty-first century AD, twelve-year-olds studying the circle are still delighted to rediscover the theorem for themselves.

Thales also discovered that the base angles of an isosceles triangle are equal. You are maybe feeling a bit dismissive. 'For goodness sake, everyone knows that,' you'll be saying. That may be so, but 2600 years ago, Thales was the *first* to know it.

Thales was also the first mathematician in history to whom specific mathematical discoveries have been attributed.

Now, one last fact about tangents and circles:

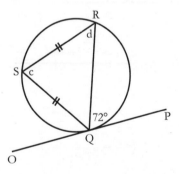

The angle between tangent and chord is equal to the angle in the opposite segment. Making this blunt statement less 'mathy', it means that ∠RQP between tangent QP and chord QR equals the angle on the other side of the chord, namely the angle marked c.,

(A chord is a line that connects any two points on a circle but the special chords that also pass through the centre of the circle are called diameters.)

$$So\ c = 72°$$

And since ΔSRQ is marked isosceles, $\angle$s SRQ and SQR=54° since the angles of a triangle add up to 180°. To complete the calculation, $\angle$SQR = 54°.

The Dreaded Decimals

The decimal point is not just a dot randomly inserted into a bunch of digits. The point is important because it separates the whole numbers from the fractions.

Let's write the headings and some numbers:

100's	10's	1's		10ths	100ths	1000ths
2	6	4	.	1	5	9

So the '1' is $\frac{1}{10}$, the '5' is $\frac{5}{100}$ and the 9 is $\frac{9}{1000}$. The further you go to the right, the smaller the fraction is.

The decimal point was introduced by Scottish mathematician John Napier (1550–1617). Until then, the above number would have looked like:

$$264{1'}{5''}{9'''}$$

(Can you imagine what a mess π to ten decimal places would have been?)

But that was by no means all that John Napier did. In 1614, after twenty years of work, he finally published *Mirifici Logarithmorum*. The entire work, with fifty-seven pages of explanation and ninety pages of tables, was in Latin, the common language among educated people at that time. Napier had invented logarithms. His log tables produced methods of calculations that changed lives all over the world and were one of the greatest advances in the history of mathematics. Log tables were only superseded by electronic calculators in the 1970s.

Before the invention of pocket calculators by Jack Kilby, Jerry Merryman and James Van Tassel, secondary-school students used log tables for multiplication, division, squares, cubes, square and cube roots, etc.

But primary-school children had first to learn decimal multiplication and division the hard way, doing sums like 396.34 × 42.7 and 70.38 ÷ 6.9 with paper and pencil only. Torture like that is now a thing of the past. Books of log tables are gathering dust in maths store cupboards. God bless Jack Kilby and his mates.

However, as I mentioned earlier, mistakes can be made when pressing little buttons with fat fingers, so every time you use the calculator, you should know what the approximate answer is. So for the above multiplication, you should think '400 times 40 = 16 000'. Now use your calculator and confirm that 16 923.718 has the point in the correct place. Likewise for the division, think: '70 ÷ 7 = 10' and then use your calculator to get 10.2.

You might think you have made some mighty blunder in a calculation like 150 ÷ 0.00003 when you get the answer 5 million. But remember, the smaller the divisor, the greater the answer. If you are still not happy, you can confirm this one without a calculator:

$$150 ÷ 0.00003$$

Move the point to the end by jumping five places.

And, to keep things equitable, attach five zeros to 150.

So you have 15 000 000 ÷ 3

= 5 million.

Rounding Off Decimals

Often, and especially when you use the calculator to divide, you end up with far too many digits after the decimal point. First, you decide how many decimal places you need. But before you discard the rest, look at the first unwanted one. If it is 0, 1, 2, 3 or 4, go ahead and discard the unwanted digits. However, if the first unwanted digit is 5, 6, 7, 8 or 9, increase the last wanted digit by 1. Examples:

(a) Round off 7.15623 to one decimal place. The second place is 5, so change the 1 to 2 and the number becomes 7.2 (1dp)

(b) Round off 45.49866 to two decimal places. The first unwanted digit is 8, so this time 49 will increase to 50 and the number becomes 45.50 (2dp)

(c) Round off 4.73128 to one decimal place. The first unwanted digit is 3, too small to bother with, so we discard it and all the other hangers-on. So the number becomes 4.7 (1dp)

Look, No Calculator!

You don't need a calculator to multiply and divide decimals by 10, 100, 1000, etc. All you do is 'let the point jump'. Examples:

(a) 42.57 × 10 (one jump →) = 425.7

(b) 7.12 × 1000 (three jumps →) and fill in the last jump with a zero = 7120

(c) 4.092 × 1 000 000 (six jumps →) and fill in the last three jumps with zeros = 4 092 000

(d) 92.5 ÷ 10 (1 jump ←) = 9.25

(e) 0.97 ÷ 100 (2 jumps ←) and fill in the last jump with a zero = 0.0097

Another Abominable Maths Joke

A mathematician met a shepherd in the hills. The shepherd said he had 59 sheep and the dog was just about to round them up. 'Oh,' said the mathematician, 'you don't need the dog. It's 60.'

Changing Negative Powers to Decimals

You learned earlier that 2^{-3} meant 1 over 2^3, getting the fraction ⅛.

In the same way, 10^{-2} means 1 over 10^2, which is ¹⁄₁₀₀, getting 0.01.

Likewise, $10^{-3} = 0.001$, $10^{-6} = 0.000001$ or one millionth. (The negative power tells you the number of zeros *including* the zero before the point.)

So that's twenty-first-century decimals. They weren't so bad after all.

Maths Joke with Apologies

What would life be like without decimals? Pointless!

Adding and Subtracting Positive and Negative Numbers

At first it is best to add and subtract positive and negative numbers on a number line which, of course, extends forever in both directions.

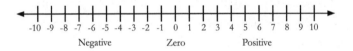

The rule is:

(1) When adding a positive number, go right.
(2) When adding a negative number, go left.
(3) When subtracting a positive number, go left.
(4) When subtracting a negative number, go right.

Let's try these:

(a) $4 + 5 = 9$ (start at 4 then 5 →)
(b) $6 - 9 = -3$ (start at 6 then 9 ←)
(c) $3 + (-4) = -1$ (start at 3 then 4 ←)
(d) $2 - (-6) = 8$ (start at 2 then 6 →)

Here, (c) and (d) may be confusing. But + (–) squashes together to make '–' so it is now like example (b); and – (–) squashes together to make '+' so it is now like example (a).

So, five more examples:

(e) $6 - 8 = -2$
(f) $-5 - (-5) = -5 + 5 = 0$
(g) $-6 - 3 = -9$
(h) $7 + (-2) = 7 - 2 = 5$
(i) $-4 + (-4) = -4 - 4 = -8$

Algebra

For many people, the move from maths with numbers to maths with letters of the alphabet is a step too far.

But you have already been doing snippets of algebra throughout this book. If you were asked how to find the circumference of a circle, you could say, 'Well, you need to know the length of the diameter and then you multiply it by a number which is approximately 3.14.' However, it would be easier to say, 'The formula is $C = \pi d$', so you are using algebra.

In algebra, and in maths generally, not only is almost every letter of our own alphabet used, but we also plunder the Greek alphabet – alpha, beta, gamma, delta . . . α, β, γ, δ . . . and the other twenty of them. They do much to enhance the mystery of maths. You are already acquainted with pi and phi.

The letters are mostly called variables because in different calculations they vary in value, but π and ϕ are called constants because they are always 3.14. . . and 1.618. . .

However, x is everyone's favourite variable, with y a poor second, and a, b and c after that. So for all maths, except arithmetic, x, the multiplication sign, is missed out altogether. If this is not possible, an asterisk * is used instead but it is surprising how little it is needed.

x IS EVERYONE'S FAVOURITE VARIABLE

So, although the formula for the area of a rectangle is Area = length times breadth, we just write A = lb. The formula for calculating the area of a kite is half the product of its diagonals, or A = ½ × d_1 × d_2, but that is written without the kisses:

$$A = ½ \, d_1 d_2.$$

Simplifying Expressions

Your expression is so simplified it's blank.

An expression is a bunch of assorted numbers and letters that need to be simplified **if possible**. For example, $4x + 6 + 3y - 2 + 8x$.

We can only simplify if we have 'like' terms. So $4x + 8x = 12x$ and $+6 - 2 = +4$ and y is on its own. So the above expression simplifies to $12x + 4 + 3y$ (not necessarily in that order as adding is commutative).

But you cannot simplify unless there are 'like' terms: $a + a^2$ cannot be simplified nor can $a + b - c + ab - bc$.

Simplify these if possible:

(a) a + 2a + 3a
(b) b + 6 + 3b – 4
(c) $2x + 2x^2$
(d) p + 3q – p – q

Answers: (a) 6a (b) 4b + 2 (c) cannot be simplified (d) 2q

Expanding Brackets

Expanding brackets means multiplying everything inside the bracket with what's outside the bracket. For example,

$$4(3x + 2y) = 4 \times 3x + 4 \times 2y = 12x + 8y$$
$$\text{and}$$
$$x(3x + 4y) = x \times 3x + x \times 4y = 3x^2 + 4xy$$

Expand the brackets in the following:

(a) $7(x - z)$
(b) $x(x + 1)$
(c) $4x(4x - 3y)$

Answers: (a) $7x - 7z$ (b) $x^2 + x$ (c) $16x^2 - 12xy$

A Silly Maths Joke

Maths test: Expand $5(x + 8)$

Student's effort: 5 $(x + 8)$

'Is this enough?'

Picking Up the Pieces

It was a sunny, but very breezy morning, and Katie decided to do the crossword in the garden. She was pondering over 4 across, 'Tease relative after time' (5 letters) when the phone rang so she dashed indoors. It was only another sales pitch to persuade her to buy solar panels but when she went back out again, she found her newspaper had blown all over the garden.

How many sheets would she have to pick up? She glanced at the first sheet. The pages were numbered 32 and 49 and 31 and 50 on the back of the sheet. The next sheet she retrieved was numbered 18 and 63 with 17 and 64 on the reverse side.

In each case the pair of numbers added to 81 so she subtracted 1. This gave her 80, the number of pages in the newspaper. Then dividing by 4 gave her the number of sheets she would have to pick up. Eventually, she retrieved all 20 sheets, put them back in order and finished the crossword.

Katie reckoned she had discovered a new algebraic formula where N is the number of sheets and P_x and P_y are the numbered pages.

$$N = (P_x + P_y - 1) \div 4$$

PS: The answer to the crossword clue is 'taunt', 'aunt + t'.

Solving Equations

$5x + 4 = 29$ is an equation and you will attempt to 'solve for x' – in other words, find the numerical value of curly x. Every equation has a left-hand side (lhs) and a right-hand side (rhs), and both sides (bs) are separated by the = sign. The sign $\therefore$ means 'therefore'.

Up in Mount Sinai, where he tarried forty days and forty nights, Moses received the Ten Commandments from God that the people of Israel had to obey.

There is an Eleventh Commandment that mathematicians have to obey: 'Thou must doeth unto the rhs of an equation what thou doeth unto the lhs.' This commandment is more useful than Moses' Tenth Commandment, the one about not coveting your neighbour's ox or donkey.

In the following examples, all the steps will be included but, as you become more confident, feel free to miss them out.

Example 1
$7x - 8 = 55$
(Add 8 to bs)
$7x - 8 + 8 = 55 + 8$
$7x = 63$
(Divide bs by 7)
$\therefore x = 9$

Example 2
$\frac{1}{3}x + 8 = 10$
(Subtract 8 from bs)
$\frac{1}{3}x + 8 - 8 = 10 - 8$
$\frac{1}{3}x = 2$
(Multiply bs by 3)
$\therefore x = 6$

Example 3
$-16 = 5x + 4$
(Change sides, we like our x's on the lhs)
$5x + 4 = -16$
(Subtract 4 from bs)
$5x + 4 - 4 = -16 - 4$
$5x = -20$
(Divide bs by 5)
$\therefore x = -4$

Example 4
8x + 7 = 2x + 43
(Subtract 2x from bs)
6x + 7 = 43
(Subtract 7 from bs)
6x = 36
∴ x = 6

Example 5
½x − 1 = ⅓x + 2
(Multiply bs by 6)
3x − 6 = 2x + 12
(Subtract 2x from bs)
x − 6 = 12
∴ x = 18

Try these:

(a) 4x + 5 = 29
(b) 7x − 15 = 3x + 5
(c) 5(3x − 1) = −50

Answers: (a) 6 (b) 5 (c) −3

However, you don't need to get answers for equations. Check your own. Just go back to the beginning and substitute your answer for x and both sides should be equal.

Algebra with a Little Added Geometry

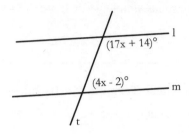

This little problem is a mixture of geometry and algebra. You did the geometry bit many pages ago. Remember? m and l are parallel lines cut by transversal t, and $(17x + 14)°$ and $(4x -2)°$ are, therefore, co-interior angles and they add up to 180°.

$$17x + 14 + 4x - 2 = 180$$
$$21x + 12 = 180$$
(Subtract 12 from bs)
$$21x = 168$$
$$\therefore x = 8$$

so the actual angles are 150° and 30°

$E = mc^2$

$E = mc^2$ is the famous equation of Albert Einstein (1879–1955). E is energy, m is mass and c^2 is the speed of light squared. The

130

speed of light is 186 000 miles per second so c^2 is an unimaginably colossal number.

The equation looks simple yet it is a concept that changed the world.

Another of These Terrible Maths Jokes

The Romans thought that algebra was easy. X was always equal to 10.

And Another One!

Dear Bereft Algebra,
Stop asking me to find your X.
She is not coming back,
Agony Aunt.

you + me = 3

Volume of Containers

Everyday containers in your home are usually square boxes, rectangular boxes and round boxes.

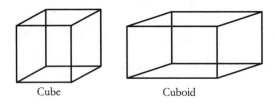

Cube   Cuboid

In maths, a square box is called a cube, a rectangular box is a cuboid and a round box is a cylinder. The formula for the volume of a cube is $V = L^3$ and the volume of the cuboid is $V = Lbh$.

The cube has a side of 10 inches so $V = 10^3 = 10 \times 10 \times 10 = 1000$ in^3. The cuboid has L = 9 inches, b = 10 inches and h = 11 inches. Now, since the dimensions are much the same as those of the cube, will the volume of the cuboid be less, more or the same? Well, since 10 + 10 + 10 = 9 + 10 + 11, maybe the volumes will be equal? However, the volume of the cuboid is $9 \times 10 \times 11 = 990$ in^3, smaller by **10** in^3. Mmm? Let's delve into this a bit more.

Cube	Cuboid	Vol. of cube	Vol. of cuboid	Difference
3 × 3 × 3	2 × 3 × 4	27	24	3
5 × 5 × 5	4 × 5 × 6	125	120	5
12 × 12 × 12	11 × 12 × 13	1728	1716	12

So, putting this into algebra:

Volume of cube = x^3

Volume of cuboid = $x(x - 1)(x + 1) = x(x^2 - 1) = x^3 - x$.

Remember the difference of two squares earlier,
where $(x^2 - 1) = (x - 1)(x + 1)$.

133

The Volume of a Cylinder

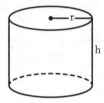

The formula for calculating the volume of a cylinder is area of base times height:

$$V = \pi r^2 h \text{ (the base is a circle with area } \pi r^2)$$

The dimensions of an empty rice pudding tin (the pudding described as 'the food of the gods') are: radius 3.7 cm and height 10 cm, making the volume $3.14 \times 3.7 \times 3.7 \times 10 = 430 \text{ cm}^3$. The can is filled with water and poured into a measuring jug and found to be 430(ish) millilitres. The measurements were not 'spot on' but they demonstrate that $1 \text{ cm}^3 = 1$ ml and for water, 1 ml weighs 1 gram.

More about Volume

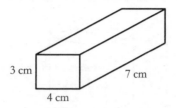

The volume of this small cuboid is $3 \times 4 \times 7 = 84$ cm³. Let's multiply each side by 10 so the new volume is $30 \times 40 \times 70 = 84\,000$ cm³. So the volume was increased a thousand-fold. Putting that into algebra, when the dimensions of a solid are multiplied by x, the volume will be multiplied by x^3.

Three-dimensional Pythagoras

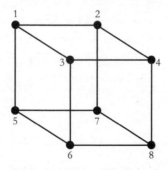

This cube shows its eight vertices. Lines joining 6 to 4, 2 to 3, 2 to 8 and nine other lines are the **face** diagonals. The lines joining 1 to 8, 2 to 6, 3 to 7 and 4 to 5 are **space** diagonals. They don't look as if they are equal in length but they are!

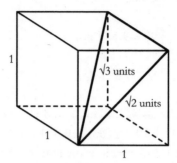

This cube has a side of 1 unit, so the length of each **face** diagonal is calculated by Pythagoras: $\sqrt{(1^2 + 1^2)} = \sqrt{2}$, the nemesis of poor Hippasus. But the length of the space diagonal is calculated by three-dimensional Pythagoras: $\sqrt{(1^2 + 1^2 + 1^2)} = \sqrt{3}$, another irrational number.

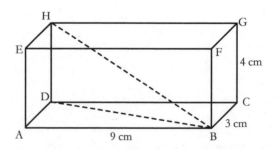

This cuboid has three pairs of equal faces: top and bottom, front and back, and left and right sides, so it will have three pairs of different face diagonals:

$BD = \sqrt{(9^2 + 3^2)} = \sqrt{90} = 9.5$ (1 dp)
$BG = \sqrt{(3^2 + 4^2)} = 5$
$AF = \sqrt{(9^2 + 4^2)} = \sqrt{97} = 9.8$ (1 dp)

The space diagonals AG, BH, CE and DF are equal and are of length $\sqrt{(9^2 + 4^2 + 3^2)} = \sqrt{106} = 10.3$ (1 dp).

All of the above results are irrational numbers except the 3, 4, 5 Pythagorean triple. It's hard to believe that Pythagoras would not admit that such numbers existed.

Foreign Currency

Foreign currency exchange rates fluctuate by small amounts daily, but sometimes over a longer period they change significantly. At the time of writing, the rate of exchange for the euro is £1 = €1.42. Six months ago, the euro rate was only £1 = €1.24. So if you are going off to Tenerife for some winter sunshine or to enjoy opera at La Scala in Milan, it is a good idea to buy your euros in advance when the rate is good.

Also, at the time of writing, the £ ↔ $ rate has been steady at £1 = $1.52, only fluctuating by a fraction of a cent per day.

But here is *a sad tale*. Archie loved skiing and he and his pals planned to go to the glittering snow playground of Vail, Colorado, to ski alongside the rich and famous on the perfect pistes, with the luxury gondolas, high-speed lifts and pretty girls. He was going to do this in style so he went to his local bank and asked for £5000 worth of dollars. The rate was £1 = $1.56 so he got 5000 × 1.56 = $7800.

A week before the holiday, he decided to improve his skiing skills in the local Scottish Cairngorm Mountains. He was bombing down a black run when he hit a mogul and went flying, his skis in one direction and his body in the other. The medics said his leg wasn't broken but it was badly sprained and there would be no ski holiday in Colorado.

He hobbled to the bank to change his dollars back to pounds. The cashier said, 'We buy dollars at 1.72. Is that OK?' Archie cheered up. £7800 times 1.72 was a lot of money!

However, he was shocked when the teller counted out £4534.88. You see, the bank was *buying* his dollars. Archie was *selling* them. So the calculation was *not* 7800 × 1.72. It was **7800 ÷ 1.72** and that was over £460 less than he had bought them for.

So the rule for all travellers, wherever you are going is:

£ → foreign money, multiply by the currency rate.
Foreign money → £, divide by the currency rate.

The rate for changing foreign money back to £s is always higher than when you buy it because the banks have to make their profit. And dividing by a higher number makes a lower answer. The advice is to spend what you have got left of your holiday money at the airport before you fly home.

Temperature

In the US, everyone has been clinging resolutely to the Fahrenheit thermometer since its invention by scientist and engineer Daniel Gabriel Fahrenheit in the early eighteenth century. The young man was also a glass-blower, giving him the perfect skills for his invention.

In his mercury thermometer, the freezing point of water is set at 32 degrees (32°F) and the boiling point of water at 212 degrees (212°F).

The rest of the world have agreed that the Fahrenheit thermometer has had its day and everyone else now uses the Celsius or Centigrade thermometer invented by Swedish scientist Anders Celsius in the 1740s. On the Celsius thermometer, 0°C is the freezing point of water and 100°C is the boiling point of water. (Some say, 'Celsius', others say, 'Centigrade'. They are essentially the same but the TV weather forecasters use Celsius, so we will go with the flow.)

In the UK, although Celsius is always used in science, business and weather forecasting, the rest of us continue to use both Celsius and Fahrenheit in the most ridiculous way. In everyday conversation, we don't even bother to attach C or F because we all know which one is meant. When you tell your pals about your touch of flu and your temperature of 100, no one imagines that your innards were bubbling at boiling point. In the summer, you say, 'Wow! It's hot! It must be in the high 60s!' (This is in Scotland, not Death Valley, California!) But in the winter, we change to the Celsius scale and complain about icy pavements and our suspicion that the temperature is at least 2 below zero. No one would ever say, 'It must be 2 under 32.'

So, if there is any need to know how to convert between Fahrenheit and Celsius scales, it is only in the UK.

The correct and accurate way to convert °C to °F is to divide by 5, multiply by 9 and add 32. For example, 30°C = (30 ÷ 5 × 9) + 32 = 86°F. To go in the reverse direction, you subtract 32, divide by 9 and multiply by 5. For example 68°F = (68 – 32) ÷ 9 × 5 = 20°C.

However, these examples have been handpicked because the calculation is easy. Try 75°F and you get involved in recurring decimals when you attempt to divide 43 by 9. It all becomes too much of a chore.

However, there is a much easier but less accurate method. To go from F to C, subtract 30 and divide by 2. For example, 60°F = (60 – 30) ÷ 2 = 15°C, nearly accurate. To go from C to F, double and add 30. Thus, 25°C = (25 × 2) + 30 = 80°F, inaccurate by 3 degrees too many.

It is useful to know that 61°F ↔ 16°C, and 82°F ↔ 28°C. And for pub quizzes, –40°F is exactly the same as –40°C. At –40, the plastic, rubber and glass of a car will shatter and exposed human skin will freeze within a minute.

What Maths Nerds Do for Fun

There are scores of ways to make 100 using all the digits in order from 1 to 9. Here are a few:

$$100 = 123 + 4 - 5 + 67 - 89$$
$$100 = 1 + 2 + 34 - 5 + 67 - 8 + 9$$
$$100 = 1 + 2 + 3 - 4 + 5 + 6 + 78 + 9$$
(and backwards)
$$100 = 98 - 76 + 54 + 3 + 21$$
If you allow × and ÷ and brackets,
$$(1 × 2 × 3 × 4) + 5 + 6 - 7 + (8 × 9) = 100$$

A MATHS NERD

Inequations

Inequations are like equations but instead of the = sign separating the left-hand side (lhs) and the right-hand side (rhs) you will see the inequality signs, <, >, ≤ or ≥. The Eleventh Commandment must be obeyed as before, along with three extra addenda.

Addendum (1)
4 < 7 but changing sides means changing signs to 7 > 4

Example (a)
$8 < x + 6$
(Change sides, change sign)
$x + 6 > 8$
$\therefore x > 2$

Example (b)
$13 > 19 - 3x$
(No need to change sides this time, just add $3x$ to bs)
$13 + 3x > 19 - 3x + 3x$
$3x + 13 > 19$
$3x + 13 - 13 > 19 - 13$
$3x > 6$
$x > 2$

Addendum (2)
−6 < −3 but dropping the negatives from both sides means 6 > 3

Example (c)
$-3x > -21$
(Drop negatives, change sign)
$3x < 21$
$\therefore x < 7$

Addendum (3)
−5 < 3 and 7 > −1 but transferring the negative to the other side means 5 > −3 and −7 < 1.

Example (d)
$4 - 5x \geq 25 - 2x$
(Subtract 4 from bs)
$4 - 5x - 4 \geq 25 - 2x - 4$
$-5x \geq 21 - 2x$
(Add 2x to bs)
$-5x + 2x \geq 21 - 2x + 2x$
$-3x \geq 21$
(Transfer − to rhs, change ≥ to ≤)
$3x \leq -21$
$\therefore x \leq -7$

Try these:

(a) $-4x - 3 > 9$
(b) $-10x - 7 < -77$
(c) $x - 4 \leq 8 + 2x$

Answers: (a) $x < -3$ (b) $x > 7$ (c) $x \geq -12$

Percentages %

Percentages are everywhere in our daily lives:

'Politicians Agree To Accept Only 2% Pay Rise'
'90% of Women Love SHINO Shampoo!'
'20% Off Your Bill at the Till'

You have to be sceptical about these statements. Two per cent added to an already large salary is a large increase. Two per cent on the minimum wage is only £5 per week extra before tax.

In the second example, did ten women get a sample of SHINO and nine of them say they liked it? (Especially since it was a free sample.) Or did 90% of five thousand women try it and write effusive letters to SHINO's head office saying how much they loved it and how SHINO would be their favourite shampoo for evermore.

And in the third example, did you notice the small print under the supermarket offer: 'when you spend £60 or more.'

So always take the percentages in adverts with a pinch of salt.

How to Calculate a Percentage of Money

To find 1% of pounds, euros or dollars, the rule is simple. Look at the 'pound' and say 'pence' and likewise for euros and dollars. For example:

$$1\% \text{ of } £60 = 60p$$
$$1\% \text{ of } €35 = 35c$$
$$1\% \text{ of } \$230 = 230c \text{ or } \$2.30.$$
So, from there, you can calculate 2%, 3%, 4%, etc.

To find 10%, look at the pounds and go back one place from the end and insert a decimal point or, if there already is a point, go back another one place. For example:

$$10\% \text{ of } £82 = £8.20$$
$$10\% \text{ of } £4.50 = £0.45$$
$$10\% \text{ of } €92.65 = €9.265 = €9.27 \text{ (2 dp)}$$
$$10\% \text{ of } €1\,000,000 = €100\,000$$
$$10\% \text{ of } \$0.65 = \$0.065 = 7c$$
$$10\% \text{ of } \$500 = \$50.$$

From there, you can calculate 15%, 20%, 30%, 40%, etc.

The connection between percentages and fractions is also a great help.

$$12\tfrac{1}{2}\% = {}^{12\frac{1}{2}}\!/_{100} = {}^{25}\!/_{200} = \tfrac{1}{8}$$
$$33\tfrac{1}{3}\% = {}^{33\frac{1}{3}}\!/_{100} = {}^{100}\!/_{300} = \tfrac{1}{3}$$

%	fr
2½	⅟₄₀
5	⅟₂₀
12½	⅛
25	¼
33⅓	⅓
66⅔	⅔
75	¾
87½	⅞

Examples

(a) 66⅔% of \$90 = ⅔ × 90 = $^{180}\!/_3$ = \$60

(b) 2½% of €80 = ⅟₄₀ × 80 = $^{80}\!/_{40}$ = €2

Percentages are commutative (remember that word?):

70% of £50 = 50% of £70 = £35

80% of £75 = 75% of £80 = ¾ × 80 = £60

64% of \$25 = 25% of \$64 = ¼ of 64 = \$16

96% of £10 = 10% of £96 = £9.60

% with the Calculator

When all else fails, there is always the calculator. However, on some calculators, you should **not** press the = sign. For 20% of

45 you press **45 × 20 %**. The answer is 9. Pressing the = sign gives 405, which is clearly not the correct answer.

On other calculators, you must press **45 × 20 % =**. With the = you get 9 but without the = you get 0.2, which is not the correct answer.

So test your own calculator to see whether it likes to do percentages with or without the = sign.

Percentage Increase and Decrease

A % increase or decrease is the difference between the original and the new data expressed as a % of the **original** data.

Example (1a)
After cutting out ice cream and chocolate biscuits from her diet, a girl's weight dropped from 60 kg to 57 kg. What was her % decrease?

$$\text{Original weight} = 60 \text{ kg}$$
$$\text{New weight} = 57 \text{ kg}$$
$$\text{Decrease} = 3 \text{ kg}$$
$$\% \text{ decrease} = \tfrac{3}{60} \times 100\%$$
$$= 5\%$$

Example (1b)
An obese man weighing 350 lb was advised by his doctor to lose weight to avoid serious illness. On his next visit to the doctor, he weighed 333 lb. What was his % decrease?

Original weight = 350 lb
New weight = 333 lb
Decrease = 17 lb
% decrease = $^{17}/_{350} \times 100\%$
= 4.9% (1dp)

So the man who lost more than twice the number of pounds than the girl did, has a lower % decrease. It seems that percentage decreases on their own provide no real information whatsoever.

Example (2a)
The manager of a factory has a rise in salary from £50 000 a year to £53 000. What is his % increase?

Original salary = £50 000
New salary = £53 000
Increase = £3000
% increase = $^{3000}/_{50\,000} \times 100\%$
= 6%

Example (2b)
A labourer in the same factory has his wage increased from £16 000 a year to £17 000. What is his % increase?

The calculation is: $^{1000}/_{16\,000} \times 100\% = 6.25\%$

Comparing % increases only, it seems that the labourer got the better deal. But did he really? Again, these calculations tell you nothing about the basic facts.

As Mark Twain, the famous nineteenth-century American author, said: 'There are lies, damned lies and statistics.'

Percentages Going Up and Percentages Coming Back Down Again

Thomas, who worked at the above factory as an accountant, bought a house for £200 000 in 2010. He was delighted when, in 2015, he had the house revalued and discovered it had increased by 15%. His house was now worth £200 000 + 15% of £200 000 = £230 000.

However, he lost his job – only the management knew why. He had to sell his house and get out of town fast. The estate agent told him he had only one offer for his house and that was 15 per cent less than the asking price. Ah well, he had broken even, Thomas thought. But he was not only a dishonest accountant, he was also a rotten mathematician.

Fifteen per cent of £230 000 is £34 500 and £230 000 – £34 500 = £195 500. He was selling his house for less than he had paid for it five years earlier.

So beware of % up and % down.

Simple and Compound Interest

Simple interest (I) is the money earned by investing a sum of money called the principal (P) at a % rate of interest (R) for a number of years (T). Simple interest is paid every year on the *original* principal only.

At the time of writing, interest rates are very low for savers.

The formula for calculating simple interest is I = PRT ÷ 100.

Example

Henry saves £1500 for two years at 4% per annum ('annum' is a Latin word meaning 'year' and is often used in financial and business transactions). His interest after two years will be 1500 × 4 × 2 ÷ 100 = £120.

If his money had been compounded at the end of the first year by adding the first year's interest on to the original principal, the calculation would be:

Year 1: I = 1500 × 4 × 1 ÷ 100 = £60
So he starts a new year with £1560.
Year 2: I = 1560 × 4 × 1 ÷ 100 = £62.40
So his total interest is £122.40 and his amount in the bank is £1622.40.

Doing these simple year-by-year calculations is fine but if Henry wanted to know what his accumulated interest would be after five years, he would have to do three more tedious calculations, starting year 3 with £1622.40. In year 3, the interest will be 1622.40 × 4 × 1 ÷ 100 = £64.90. In year 4, the interest will be 1687.30 × 4 × 1 ÷ 100 = £67.49 and lastly, in year 5, 1754.79 × 4 × 1 ÷ 100 = £70.16.

So, once again blessing Jack Kilby and his fellow pocket calculator inventors, Henry's accumulated wealth over five years is £1500 + £60 + £62.40 + £64.90 + £67.49 + £70.16 = £1824.95.

However, this lengthy, boring calculation can be done in seconds using a wonderful formula for compound interest. It is $A = P(1 + r/100)^t$ where A is the final amount.

For the above example:

$$A = 1500(1 + 4/100)^5$$
$$= 1500(1 + 0.04)^5$$
$$= 1500 \times 1.04^5$$
(on your calculator, press 1.04, x^y, 5 =)
$$= 1500 \times 1.21665$$
$$= £1824.97$$

Apologies for the extra 2p. It was caused by rounding up the value of 1.04^5. Probably Henry never noticed.

The Rule of 72

This rule with its simple formula, $T = 72/r$, is an easy and reasonably accurate way of finding out how long any sum of money invested at a rate of r% will take to double. There is an exact formula but it needs complicated maths to use it. Even the professional money-men prefer the Rule of 72 for everyday usage.

For example, how many years will it take for £1000 to double at a rate of 4% per annum? The sum of money is irrelevant to the calculation and the answer is simply 72 ÷ 4 = 18 years. So Henry will have to wait eighteen years for his £1500 to become £3000.

In most cases, the Rule of 72 is 'out' by less than three months.

Here is part of a table comparing the exact calculation in years compared with the Rule of 72:

% rate	Exact	72 rule
3	23.45	24
4	17.67	18
5	14.21	14.4
6	11.9	12
7	10.245	10.286
8	9.006	9
9	8.043	8

To change the decimal part to months, multiply by 12. For example, for the 7% rate, 0.245 years = 0.245 × 12 = 3 months and 0.286 years = 0.286 × 12 = 3½ months.

Checking the Rule of 72

Henry reckoned it would take eighteen years to double his money when he invested £1500 at 4%. Let's check it with the $A = P(1 + \%_{100})^t$ formula.

$$A = 1500(1 + 0.04)^{18}$$
$$= 1500(1.04)^{18}$$
(Get the calculator out!)
$$1.04^{18} = 2.0258 \text{ (press 1.04, } x^y, 18, =)$$
$$A = 1500 \times 2.0258$$
$$= £3038.70$$

Henry will be delighted that his investment has doubled after eighteen years with an extra little bonus of £38.70.

Prime Numbers

Prime numbers have fascinated mathematicians for at least 2500 years, the ancient Greeks being given the credit (or blame) for starting an obsession that has never faltered throughout the centuries.

But first of all, what are prime numbers? They are special numbers that cannot be divided by any number except themselves and 1, and they stretch along the number line forever.

If numbers have souls – and Pythagoras believed that they did – then 1 would be the most unhappy little soul because it is not the leader of the greatest and most famous infinite sequence of numbers in mathematics. The mathematical police do not allow 1 to be a prime number as it does not fulfil both of the above conditions. It is probably no consolation that 1 leads almost every other number sequence.

So 2 is the first prime number and the only prime that's even. Here are the numbers 1 to 100. The prime numbers are in bold italic type. There are twenty-five of them.

1	*2*	*3*	4	*5*	6	*7*	8	9	10
11	12	*13*	14	15	16	*17*	18	*19*	20
21	22	*23*	24	25	26	27	28	*29*	30
31	32	33	34	35	36	*37*	38	39	40
41	42	*43*	44	45	46	*47*	48	49	50
51	52	*53*	54	55	56	57	58	*59*	60
61	62	63	64	65	66	*67*	68	69	70
71	72	*73*	74	75	76	77	78	*79*	80
81	82	*83*	84	85	86	87	88	*89*	90
91	92	93	94	95	96	*97*	98	99	100

This grid of prime numbers is called the Sieve of Eratosthenes. Eratosthenes lived in Greece in the third century BC, a time of great advancement in learning. An all-round genius, he was a mathematician, poet, astronomer, geographer and music theorist. But his nickname was Beta, the second letter in the Greek alphabet. His teacher and friend Archimedes was always the alpha genius. (Archimedes has a crater on the moon named after him, whereas Eratosthenes does not. However, Eratosthenes does have his name on a minor planet.)

Remember, the above grid shows only the first twenty-five prime numbers. They stretch randomly along the number line forever.

Look at the twenty-five prime numbers. Each one, apart from 2 and 3, lies before or after a number on the 6 times table.

This is true for every prime number along the number line. **BUT!** This is not an exclusive fact for prime numbers only. As you will see, 35 and 91 (both of which divide by 7), stand next to 36 and 90 respectively. So 6n ± 1 is *not* a formula for finding prime numbers; only a first check that they *might* be prime.

In fact, there is no formula for finding primes *exclusively* or for proving that a number is prime. There are many formulae like the simple '4n + 3' that yields 4 × 1 + 3 = 7 and 4 × 2 + 3 = 11 and 4 × 5 + 3 = 23, and many more, but this fails at 4 × 3 + 3 because 15 is not prime.

For ordinary mortals, the only way to prove a number is prime is by the method of hard slog – dividing a potential prime by other primes until all possibilities are exhausted. For example, is 617 a prime number? If you remember the divisibility tests, you will see that 3, 5, 7 and 11 are not divisors but you would need to try 13, 17, 19 and 23. In this case, you don't need to go further because 29 × 29 is more than 617 so 617 is definitely prime. There are now other ways of checking that a number is prime other than the hard slog method but none of them are quick and easy.

Mathematicians throughout the centuries have enjoyed looking for new prime numbers but this has become a mathematical sport in the twenty-first century. Using powerful computers, IT geeks have been discovering mega-multi-maths Olympic Gold prime numbers with millions of digits. Why do they do it? Probably, just because they can.

Very large prime numbers are now used in encryption

software and technology but do they need them that big? They appear to have no use whatsoever. But, like the 'useless' binary numbers of Gottfried Leibniz, maybe someday they will be very important.

Marin Mersenne

Marin Mersenne was a seventeenth-century French priest and mathematician. In his monastery in Paris, he spent years studying prime numbers. You have to admire him. He had none of the luxuries of present-day mathematicians. He worked in his cold monastic cell by candlelight with only paper and quill, and yet his formula for finding prime numbers is still being used in the twenty-first century. The formula is simple:

$$P = 2^p - 1$$

where P is the newly discovered prime number and p is an already known smaller prime number. For example $2^3 - 1$ gives 7 and getting more ambitious, $2^{19} - 1 = 524\,287$. (Try that without a calculator!) But, unfortunately, the formula fails for some prime number powers. For example, $2^{11} - 1 = 2047$ and this number divides by 23 so 2047 is not prime.

However, continuing with the formula, Leonhard Euler, the renowned eighteenth-century Swiss mathematician, found $2^{31} - 1 = 2\,147\,483\,647$ and what's more he proved it was prime by the hard slog method!

Euler said, 'Mathematicians have tried in vain to discover some order in the sequence of prime numbers and we have reason to believe that it is a mystery into which the human mind will never penetrate.' Nearly three hundred years later, nothing has changed.

Mind the Gap

Mathematicians love to see pattern and order in their work and that is what they cannot find in prime numbers.

There is certainly no pattern in the gaps between prime numbers. For example, 521 and 523 are next door to each other but the next one along after that is 541. Much further along the number line 360 287 and 360 289 are prime pals but the next prime after 360 653 is 360 749, with a gap of ninety-five composite numbers between them. (In maths, 'composite' is the fancy name given to numbers that are not prime.)

9109: A Well-behaved Prime

However, some individual prime numbers are ever so well be-haved and are full of pattern and order.

$$9109 \times 1 = 09\,109 \rightarrow 19$$
$$9109 \times 2 = 18\,218 \rightarrow 20$$
$$9109 \times 3 = 27\,327 \rightarrow 21$$
$$9109 \times 4 = 36\,436 \rightarrow 22$$
$$9109 \times 5 = 45\,545 \rightarrow 23$$

$$9109 \times 6 = 54\,654 \rightarrow 24$$
$$9109 \times 7 = 63\,763 \rightarrow 25$$
$$9109 \times 8 = 72\,872 \rightarrow 26$$
$$9109 \times 9 = 81\,981 \rightarrow 27$$

Admire how the columns run up and down from 1 to 9.

Circular Primes

A circular prime is one which remains prime when the digits are in another order. For example 71, 37, 113, 1193 and 19 937. Circular primes can only consist of the digits 1, 3, 7 and 9 as 0, 2, 4, 5, 6 and 8 at the end of a number makes it a composite number.

Circular primes above five digits are rare but mathematicians enjoy searching for them.

Sexy Prime Pairs

'Great!' you'll be thinking. 'Some good fun at last!' Sorry! This is not X-rated stuff, it is just another of those mathematicians' terrible jokes.

161

You see, '*sex*' is the Latin prefix for 6. The mathematicians could have chosen the Greek prefix for 6, which is '*hex*' as they did with hexagons. But then, hexy prime pairs wouldn't have been uproariously funny, would they?

Sexy prime pairs are merely two adjacent prime numbers that differ by 6, like (23 and 29), (31 and 37), and another forty or so under 500.

There are sexy prime threesomes and foursomes. At the beginning of the last century the years 1901, 1907 and 1913 were sexy prime trios and the century ended with 1987, 1993 and 1999.

A sexy prime foursome you will remember from the Sieve of Eratosthenes, a few pages back, is (61, 67, 73, 79). In about 1280 years from now, the sexy prime foursome (3301, 3307, 3313, 3319) will herald in the thirty-fourth century.

Will anyone be bothering about prime numbers then?

Polygons

What is a polygon? No! No! It is not a dead parrot! That is one silly maths joke you have been spared. The word has a Greek derivative and means a many-sided shape. A regular polygon has equal sides and equal angles. Polygons have names according to their number of sides and angles.

The Pentagon

To draw the pentagon and any of the following polygons, draw a circle with a compass, not less than radius 5 cm for accuracy. Then divide 360° by the number of sides in your selected polygon. For the pentagon overleaf, the centre angle is 72°. Choose a good starting place, this time at centre top, and use a protractor to mark little notches at 72° around the circumference of your circle. Join the notches and tidy up your drawing.

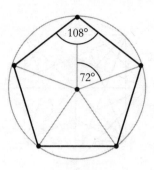

Pentagon

Having learned how to draw polygons, school students love to draw 'fancy' ones and colour them in.

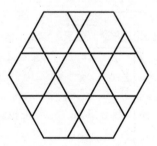

Hexagon

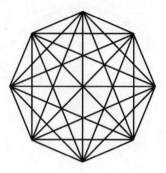

Octagon

The nonagon (nine sides) and decagon (ten sides) are equally easy to draw but the heptagon is less accurate as the answer to 360° ÷ 7 is one of those never-ending irrational numbers.

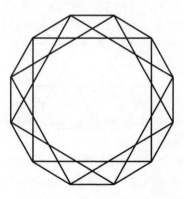

The twelve-sided regular dodecagon

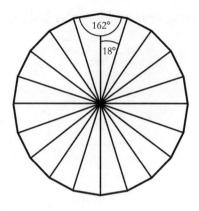

And a twenty-sided regular icosagon

The formula to find the sum of the angles of a polygon is $S = 180(n - 2)°$ where n is the number of sides. For the pentagon it is $180 × (5 - 2) = 540°$ with each angle $540 ÷ 5 = 108°$. For the octagon, it is $180 × (8 - 2) = 1080°$ with each angle $1080 ÷ 8 = 135°$. For the icosagon, it is $180 × (20 - 2) = 3240°$ with each angle $3240 ÷ 20 = 162°$.

So the more the sides of a polygon, the bigger the angle.

At first glance, the icosagon above looks like a circle and if it were possible for us to draw a hundred-sided polygon, it would be even more like a circle.

Based on that observation, Archimedes discovered the formula for the area of a circle over 2200 years ago.

Pythagoras and Regular Polygons

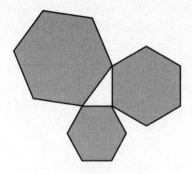

Could the Theorem of Pythagoras extend to polygons? Could it be that the area of a regular hexagon on the hypotenuse of a right-angled triangle is equal to the sum of the areas of the regular hexagons on the other two sides?

Let's say that the triangle is the 5, 12, 13 Pythagorean triple. There are several ways of calculating the area of a hexagon depending on the data given. But we only know the length of the sides and an easy formula is A = 2.598 times the square of the side length.

So, for the largest hexagon, A = 2.598 × 13² = 439.06. And the sum of the areas of the other two hexagons = 374.11 + 64.95 = 439.06.

In fact, the extended version of the famous theorem works for any regular shape that can be built on to the sides of the right-angled triangle.

EASY AS PI

How to Be a Clever Clogs: A Maths Trick

Ask any friend to write down any three numbers from 1 to 9 and keep them hidden from you. Give them the following instructions:

Double the first number, add 5 and then multiply by 5.
Add on the second number and multiply by 10.
Finally, add on the third number.
Ask them what the result of their calculation is.
You subtract 250.
Now tell them what their original numbers were.
Don't tell them how it's done.
If they chose 3, 1 and 8, the calculation would be:

$$(3 \times 2 + 5, \text{ then } \times 5) = 55$$
$$(55 + 1) \times 10 = 560$$
$$560 + 8 = 568$$
(Now subtract 250)
$$568 - 250$$
$$= 318$$
Hey presto, 3, 1 and 8!

Fermat's Last Theorem

The Theorem of Pythagoras gave us $a^2 + b^2 = c^2$ and an infinite number of Pythagorean triples.

But that's it! The $a^n + b^n = c^n$ formula goes no further than $n = 2$. There is no $a^3 + b^3 = c^3$ or $a^4 + b^4 = c^4$ or any other power of n.

This was what Pierre de Fermat wrote in 1637 in the margin of his favourite book, *Arithmatica*, by the third-century AD Greek mathematician Diophantus. Fermat was a French lawyer but he was far more interested in maths than his chosen profession. He apologised for not including the proof as there was not enough space for it in the margin beside his proposition.

His proof was never found. So strictly speaking it was not Fermat's Last Theorem, it was merely Fermat's Last Conjecture. And really, it was not that interesting. He was telling us what he thought was not true, unlike Pythagoras who gave us a positive, verified mathematical statement. However, it became a point of honour for almost every mathematician to attempt a proof. That did not include Carl Friedrich Gauss who, two centuries later, dismissed the proposition as being of little interest. Gauss is regarded as one of the world's best number theorists. You will remember reading about him earlier in the book. If Gauss had attempted the proof, he might have been successful.

Mathematicians searched for counter examples to disprove Fermat's Theorem. One almost succeeded. He thought he had discovered that $6^3 + 8^3 = 9^3$. But it isn't! $6^3 + 8^3 = 216 + 512 = 728$ and $9^3 = 729$, so, unfortunately, he was so near yet so far!

Strangely, it caught the imagination of many non-mathematicians, people who had never squared or cubed a number since they left school. Fermat's Last Theorem has appeared in the media, including *Dr Who*, *Star Trek*, *The Simpsons* and, more recently, in *The Girl Who Played with Fire*, the second novel in the Millenium trilogy by Stieg Larsson.

Then, 358 years after the discovery of that scribbled proposition in an old book, the theorem was proved at last. To worldwide acclamation, Sir Andrew Wiles published his proof in 1995. The professor's work filled 130 pages.

And there was Fermat apologising that he couldn't squeeze the proof into the margin of his book!

Simultaneous Equations

For once, the words describe the maths. We are going to solve two equations simultaneously or at the same time.

Here is a simple problem. Two numbers add to 16 but when you subtract them, you get 10. What are the numbers? You could attempt a trial-and-error method but it would probably take you ages. It's easier to do some algebra.

$$a + b = 16$$
$$a - b = 10$$

These two equations are exactly as you want them because you are always going to aim for '**equal in number but opposite in sign**'. And there they are, +b and −b, so you are happy. Now add the two equations, getting $2a = 26$ (the b's have disappeared).

So $a = 13$ and glancing at the top or bottom equation, you can see that b must be 3.

Example (a)

$$3x - 2y = 20$$
$$5x + 2y = 44$$

This one is just as easy because you have a pair of '**equal in number but opposite in sign**' y's. Add the equations.

$$8x = 64 \text{ (the y's have disappeared)}$$
$$\therefore x = 8$$

Now choose the equation you like most, probably the second one, and, substituting $x = 8$, you have $40 + 2y = 44$, so $y = 2$. You have simultaneously calculated that $x = 8$ and $y = 2$.

But life isn't always that easy. For the next example, you will also have to apply that Eleventh Commandment again: 'What thou doeth unto the lhs of an equation thou must doeth unto the rhs.'

Example (b)

$$2a - b = 10$$
$$3a + 2b = 29$$

Glancing at the b's, you see they are opposite in sign but not equal in number. So, obeying the Eleventh Commandment, multiply all the top line by 2. The equations are now:

$$4a - 2b = 20$$
$$3a + 2b = 29$$
(Add these together and the b's disappear)
$$7a = 49$$
$$\therefore a = 7$$

(Now choose the easier equation, 2a − b = 10)
$$14 - b = 10$$
$$\therefore b = 4$$

Example (c)
In the next example, things are getting harder:

$$3x - 8y = 12$$
$$x - 5y = -3$$

This time you don't have **equal in number or opposite in sign**. You have a choice but it is easier to multiply the bottom line by 3 and change the signs.

So the two equations are now:

$$3x - 8y = 12$$
$$-3x + 15y = 9$$
(Add these together and the *x*'s disappear)
$$7y = 21$$
$$\therefore y = 3$$
(Return to the top equation, and substitute y = 3)
$$3x - 24 = 12$$
(Add 24 to bs)
$$3x - 24 + 24 = 12 + 24$$
$$3x = 36$$
$$\therefore x = 12$$
(Now check that *x* = 12 and y = 3 fit both equations)

Example (d)

Lastly, this one will be done by two methods and you can decide which is easier.

$$4a + 3b = -10$$
$$2a - 5b = -18$$

(Double the bottom line and change the signs)

$$-4a + 10b = 36$$

(Add the first and third lines)

$$13b = 26$$
$$\therefore b = 2$$

(Substitute b = 2 in line 2)

$$2a - 10 = -18$$

(Add 10 to bs)

$$2a - 10 + 10 = -18 + 10$$
$$2a = -8$$
$$\therefore a = -4 \text{ and } b = 2$$

OR

$$4a + 3b = -10$$
$$2a - 5b = -18$$

(The LCM of 3 and 5 = 15.
Multiply the first line by 5 and the second line by 3)

$$20a + 15b = -50$$
$$6a - 15b = -54$$

(So now the b's are equal but opposite in sign)

(Add)

$$26a = -104$$

$$\therefore a = -4$$

(Substitute a = -4 in the first line)

$$-16 + 3b = -10$$

(Add 16 to bs)

$$-16 + 3b + 16 = -10 + 16$$

$$3b = 6$$

$$\therefore b = 2 \text{ and } a = -4$$

Try these yourself:

(a) $2a + b = 7$
    $3a - b = 8$
(b) $a - 2b = 1$
    $2a - 3b = 5$
(c) $2a + 3b = 26$
    $4a - 6b = 28$

Answers: (a) a = 3, b = 1 (b) a = 7, b = 3 (c) a = 10, b = 2

Mathematical Mind Reading: A Maths Trick

Here is a number trick you could try out on young friends. Write down any five-digit number. To avoid the 'borrowing' business, you may suggest nice big numbers, like 57 989. But if they are little number-whizz-kids, they can choose any digits they fancy.

Add the digits and subtract. So you have:

$$\begin{array}{r} 57\,989 \\ 38 \\ \hline \mathbf{57\,951} \end{array}$$

Now ask your friend to cross out any digit but keep it a secret. Suppose '7' is crossed out. Ask them to add the remaining numbers: 5 + 9 + 5 + 1 = 20. Then you subtract 20 from the next multiple of 9, which is 27. So, 27 – 20 = 7, and 7 is the number your friend crossed out. It always works, even for awkward little customers who choose to start with 50 000 or 11 111.

The Reciprocals of Prime Numbers

Once again, some mathematician has taken a word from the English language and given it a new meaning. Reciprocal is an adjective and is often used before 'agreement', meaning an arrangement mutually agreeable to both parties. In maths, reciprocal is a noun and simply means '1 over'. So the reciprocal of 2 is ½ = 0.5, and the reciprocal of 5 is ⅕ = 0.2. In fact, the only numbers that have an exact decimal reciprocal are those whose sole divisors are 2 and/or 5. For example, the reciprocal of 125 (5 × 5 × 5) = ¹⁄₁₂₅ = 0.0008 exactly and the reciprocal of 40 (2 × 2 × 2 × 5) = ¹⁄₄₀ = 0.025 exactly. The decimal reciprocals of all other numbers go on forever.

Apart from the reciprocals of the first three prime numbers, 2, 3 and 5, the reciprocals of prime numbers are very interesting. Well, that is, of course, the prime numbers we can cope with. When recently discovered super-mega primes with millions of digits are studied, would their reciprocals be calculated also? And how many digits would they have?

However, the reciprocal of 7 is nicely manageable. It is: 0.142857/142857/142857/142857 . . . look, it repeats itself

177

after six digits. This repeated chunk is called the repetend. Let's study this number 142 857 and you will notice:

(a) The digits add to 27, a multiple of 9.
(b) It divides by 9.
(c) Split the number in half and 142 + 857 = 999.
(d) Split the number in thirds and 14 + 28 + 57 = 99.

Now let's do some multiplying:

$$142\,857 \times 1 = 142\,857$$
$$142\,857 \times 2 = 285\,714$$
$$142\,857 \times 3 = 428\,571$$
$$142\,857 \times 4 = 571\,428$$
$$142\,857 \times 5 = 714\,285$$
$$142\,857 \times 6 = 857\,142$$

This little family of numbers is immaculate. There is never a number out of place. And not only do the numbers in each row add up to 27, the same is also true for the columns.

However, when you multiply 142 857 by 7, the pattern disappears and $142\,857 \times 7 = 999\,999$.

But there is one more surprise. $142\,857^2 = 20\,408\,122\,449$, and if you split that huge number into 20 408 and 122 449 and add:

$$\begin{array}{r} 20\,408 \\ 122\,449 \\ \hline 142\,857 \end{array}$$

Isn't that simply mind-blowing?

Pattern with 142 857

Draw a circle and lightly draw a radius from the centre to the top. With a protractor, mark angles of 40° round the circumference and number them from 1 to 9. Starting at **1** draw lines from **1 → 4 → 2 → 8 → 5 → 7** and back to **1**. Then complete the triangle, **3 → 6 → 9** and back to **3**. Tidy up your drawing.

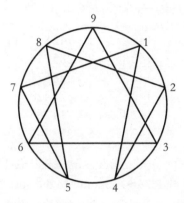

A Few More Prime Number Reciprocals

The reciprocal of 11 is $\frac{1}{11}$ = 0.09090909. . . and it isn't very interesting, but note that the repetend is 09.

The reciprocal of 13 is 0.076923/076923/07 . . . and the repetend is 076923 and the digits add to 27. It has some modified magic.

$$076923 \times 1 = 076923$$
$$076923 \times 10 = 769230$$
$$076923 \times 9 = 692307$$
$$076923 \times 12 = 923076$$
$$076923 \times 3 = 230769$$
$$076923 \times 4 = 307692$$

The 076923 pattern is there horizontally and vertically but the multipliers had to be especially chosen. But, like the reciprocal of 7, if you split 076923 in the middle and add, you get 076 + 923 = 999; and split it in thirds: 07 + 69 + 23 = 99.

The reciprocal for 17 goes the full sixteen decimal places, 0.05882352↔94117647, and again if you do the split at ↔, you get 99999999.

The number 19 has a great reciprocal but it goes for the maximum eighteen decimal places: $\frac{1}{19}$ = 0.052631578947368421 and you can multiply 052631578947368421 by any number from 1 to 18 just by finding the correct starting place, reading off the digits until you get to the end, then circling back to the beginning and stopping where you first started. It is easy to find the starting place. Suppose you want to multiply by 7, start with the number 3 after the 7. If you want to multiply by 17 you choose to start with 8, the larger number after the 7.

This only works for numbers 1 to 18. If you try multiplying by 19 all you get is a swarm of 9's.

Again, the digits of the repetend add to 81, a multiple of 9, and the number can be split in half; and guess what the two halves, 052631578 and 947368421, add to?

(A certain teacher, now retired, had the reciprocal of 19 in huge figures on the back wall of her classroom. Her students were amazed at her ability to multiply the 18-digit number by any number up to and including 18 in her head – until one little spoilsport asked why she always needed to put on her glasses to do the calculation.)

As the prime numbers get larger, in most cases the reciprocals get larger and go their full length, which is 1 less than the prime number itself. But the reciprocal of 73 repeats after eight decimal places. It is 0.01369863 . . . with the sum of the digits 36, again a multiple of 9. The digits split to 0136 and 9863, adding to 9999.

Returning to the multimillion prime numbers and the geniuses who discovered them – if they *did* calculate the reciprocals, did they then check if the digits added to a multiple of 9 and whether there was a repetend? If so, did they split it and what did they find?

I don't suppose we will ever know.

Perfect Numbers

A mathematician has a different concept of perfection from everyone else on the planet. To normal people, a baby's skin or a tiny blue tit just out of the nesting box or Mozart's Piano Concerto No. 21 are perfect.

But a mathematician gets all excited about perfect numbers, the first of which is 6. You see the divisors of 6, not including 6 itself, are 1, 2 and 3, and these add up to 6. You might remember that 6 is also the third triangular number.

Perhaps you are somewhat underwhelmed and not yelling, 'Wow!' But getting slightly more exciting, the factors of 28 are 1, 2, 4, 7, 14 and they add up to 28, and what's more, 28 is the seventh triangular number ($7 \times 8 \div 2 = 28$). There isn't another perfect number until 496, with its divisors 1, 2, 4, 8, 16, 31, 62,

124 and 248, adding to 496, and again it is a triangular number $(31 \times 32 \div 2 = 496)$.

The next one is 8128, with divisors 1, 2, 4, 8, 16, 32, 64, 127, 254, 508, 1016, 2032 and 4064. They add up to 8128, honest! It is also the 127th triangular. (Do a 'Gauss' again: $127 \times 128 \div 2 = 8128$.)

The four above were found by Euclid, the famous fourth-century BC Greek mathematician you learned about earlier. Obviously he had time to think about perfect numbers as well as all that geometry.

The fifth one, 33 550 336, was found *eighteen centuries later* in a medieval manuscript. The mathematician is unknown but he must have been a genius. Just imagine someone, with no calculating devices whatsoever, discovering the divisors of that mega-number and adding them! And did he know that the number was the 8191st triangular number? Remember though, all perfect numbers are triangular numbers but only a few triangular numbers are perfect. It's like saying, 'All buttercups are yellow flowers but not all yellow flowers are buttercups.'

After the fifth perfect number, there was approximately one new perfect number discovered per century, and in spite of the advent of powerful computers, perfect numbers are still not appearing thick and fast.

All of them so far have been of the form $2^{n-1}(2^n - 1)$. Use your calculator to check the fourth one: $2^6(2^7 - 1) = 64(128 - 1) = 64 \times 127 = 8128$. But this formula also yields an infinite number of non-perfect numbers like $2^3(2^4 - 1) = 8(16 - 1) = 120$, which is not a perfect number.

So far, fewer than fifty perfect numbers have been found. After the fifth one, they became monsters. The latest one, recorded in 2013, written out would have nearly 34 million digits! There may be others a very long way off in the number line that may never be found.

René Descartes, the renowned seventeenth-century mathematician and philosopher, said, 'Perfect numbers, like perfect men, are very rare.'

Denmark and Dog: A Maths Trick

Ask a young friend to choose a number from 1 to 9 but not to reveal their choice. Then ask them to multiply their number by 9 and add the digits. Next, subtract 5 and turn this number into a letter of the alphabet. With this letter, ask your friend to name a country and an animal. They will most likely say Denmark and dog.

Explanation
Chosen number = 7
$$7 \times 9 = 63$$
$$6 + 3 = 9$$
$$9 - 5 = 4$$
$$4 = D$$

(Whatever the chosen number, this number is always 4 because the digits of a number on the 9 times table always add to 9.)

It's fun to do this with a class of young students. One of them might just come up with deer or donkey and maybe even Djibouti!

Ratio

Example 1

You find an easy recipe for making chocolate crispies. Just melt some chocolate in a bowl over simmering water, take it off the heat and stir in Rice Krispies. Then you spoon the mixture into paper cases and leave the cakes to harden.

The ingredients required are 420 g of chocolate and 315 g of Rice Krispies. However, there is only a 200 g bar of chocolate in the cupboard, although there is a whole box of Rice Krispies. If the shops are closed and you can't get more chocolate, what weight of Rice Krispies will you need?

You write the required ingredients as a ratio with a colon in between:

$$C : R$$
$$= 420 : 315$$
$$= 4 : 3$$

(dividing by 105 or 3, 5, and 7 in stages)

Now you have 200 g of chocolate and you need x g of Rice Krispies.

So changing the ratio to a fraction,
$$\tfrac{4}{3} = \tfrac{200}{x}$$

then cross multiply

$$4x = 600$$
$$\therefore x = 150$$

So for 200 g of chocolate, you need 150 g of Rice Krispies.

Example 2
A barman's recipe for a 240 ml glass of vodka and orange cocktail is three parts orange juice to one part vodka. How much of each will he pour?

$$J : V$$
$$= 3 : 1 \text{ (4 parts in total)}$$
So ¾ will be orange juice and ¼ vodka
$$\therefore ¾ \text{ of } 240 = 180 \text{ ml of orange juice}$$
and ¼ of 240 = 60 ml of vodka.

Ratios behave like fractions. A ratio of 2 : 5 is the same as ²⁄₇ and ⁵⁄₇, and a ratio of 9 : 1 is the same as ⁹⁄₁₀ and ¹⁄₁₀.

Similar Triangles

The word 'similar' in everyday life means 'much the same without being identical'. You decide not to buy a blue sweater because you already have a similar one.

But in maths, similar triangles are identical in shape but not in size. They are scaled-up or scaled-down versions of each other. What makes them identical in shape are their equal angles.

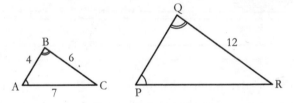

In the diagram above, angles A, B and C are equal to angles P, Q and R. Therefore their sides will be in the same ratio. QR : BC = 12 : 6 = 2 : 1, so sides PQ and PR are 8 and 14.

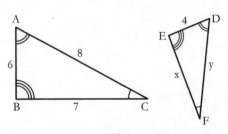

In the previous diagram, the triangles are also similar although they are not sitting tidily side by side. So $x : 7 = 4 : 6$. Turning this into an equation, $\frac{x}{7} = \frac{4}{6}$. Cross multiply and $6x = 28$ so $x = 28 \div 6 = 4\frac{2}{3}$. Also, $y : 8 = 4 : 6$, so $\frac{y}{8} = \frac{4}{6}$. Cross multiply and $6y = 32$ so $y = 5\frac{1}{3}$.

How to Find the Height of a Tree Using Similar Triangles

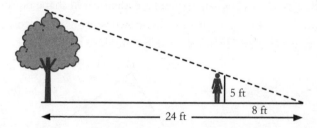

On a sunny day, find the length of your friend's shadow using a tape measure. Then she can help you measure the shadow of the tree. If x is the height of the tree, then:

$$x : 24 = 5 : 8$$
$$\text{so } \frac{x}{24} = \frac{5}{8}$$
$$\text{Cross multiply}$$
$$8x = 120$$
$$x = 15$$

So the height of the tree is 15 ft.

The Great Pyramid of Giza

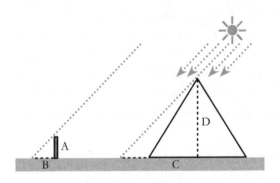

Thales of Miletus, who gave us the very first theorem, loved to travel. One of his tours took him to the Great Pyramid of Giza in the vast desert near modern-day Cairo. It was built for Khufu, the Fourth Dynasty Pharaoh about 4600 years ago. Of the Seven Wonders of the Ancient World, it's the only one that remains relatively intact. This amazing feat of engineering and construction was already 2000 years old when Thales saw it, 2600 years ago.

He asked how high it was but no one knew. So Thales decided to find out for himself. First, he measured one side of the square base, and divided by 2 (C). Then he planted a stick into the sand and waited until the time of day when the stick (A) and its shadow (B) were the same length, thus creating an isosceles right-angled triangle. So, he reckoned, the vertical height of the pyramid (D) would be equal to half the length of the base plus the length of the shadow. When the lengths of the stick and its shadow were equal, he quickly measured the length of the pyramid's shadow and added it to (C). So he was able to tell his travelling companions how high the pyramid was.

We now know that the height of the great pyramid was 481 feet and the length of the base 756 feet, but in recent years it has lost part of its top. We don't know which units Thales used for measuring. It may have been cubits or simply just paces.

Gradient

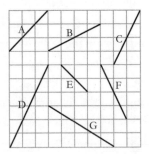

The gradient or slope of a line is measured by calculating vertical units divided by horizontal units. The units can be metric or imperial, or simply, as above, squares on graph paper. If the line slopes backwards, it will have a negative gradient.

In maths, we use the symbol m for gradient.

In the diagram above:

(A) m = 3 ÷ 3 = 1 or 45⁰

(B) m = 2 ÷ 4 = ½

(C) m = 4 ÷ 2 = 2

(D) m = 6 ÷ 3 = 2

(E) m = 2 ÷ 2 = –1

(F) m = 4 ÷ 2 = – 2

(G) m = – ⅗

E, F and G have a negative gradient because they slope backwards.

A horizontal line has no slope so m = 0. You cannot calculate the gradient of a vertical line because m would be $x \div 0$ which, as you know, is impossible. So the gradient of a vertical line is, as mathematicians like to say, undefined.

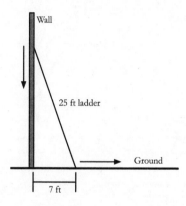

To calculate how far up the ladder in the diagram above reaches, we use old Pythagoras again:

$$7^2 + h^2 = 25^2$$
$$49 + h^2 = 625$$
$$h^2 = 625 - 49$$
$$= 676$$
$$h = \sqrt{676}$$
$$= 24$$

193

The gradient of the ladder is $^{24}/_7 = 3.43(2dp)$, but for ladders against walls it is more common to state the size of the angle the ladder makes with the ground. Mathematicians call it 'the angle of elevation'. The ladder in the diagram is roughly drawn to scale and the angle is 72^0(ish).

In the trigonometry chapter later, you will learn how to calculate this angle.

The Gradient of Perpendicular Lines

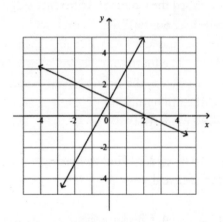

For the backward-sloping line, $m_1 = -\,^3/_6 = -\frac{1}{2}$. For the forward sloping line, $m_2 = \,^8/_4 = 2$, and notice the lines are perpendicular to each other and $2 \times -\frac{1}{2} = -1$. So when two lines are perpendicular to each other, $m_1 \times m_2 = -1$, or, in your best algebra, $m_1 m_2 = -1$.

Number Sequences

The first number sequence we will look at is the Arithmetic Progression, usually abbreviated to AP. An AP is a line of numbers like 5, 9, 13, 17, 21, 25, 29 . . . which, of course, goes on forever.

You will notice that every pair of numbers differs by 4 and this number is called the Common Difference (d). The starting number is called 'a'. So, for this AP, a = 5 and d = 4. The AP could be written out as a, a + d, a + 2d, a + 3d, a + 4d, a + 5d, a + 6d, a + 7d . . . The fourth number (or 'term' as mathematicians call it) is 'a + **3**d', so the eightieth term would be 'a + **79**d' and in general the nth term is 'a + (n – 1)d'.

Here are three more APs:

(a) 7, 10, 13, 16, 19, 22. . . Find the twentieth term. This will be 'a + 19d' where d = 3.

So 7 + 19 × 3 = 64. Check the answer by writing all twenty terms.

(b) 6, 2, –2, –6, –10, –14, –18. . . Find the twelfth term. This will be 'a + 11d' where d = –4. So 6 + (11 × –4) = 6 + (–44) = –38. Again check the answer by writing all twelve terms.

(c) On his sixth birthday, Bruce's mum says he can have £1 weekly pocket money with an increase of 5p every week.

Writing down Bruce's pocket money in pence as the weeks go by, it will be: 100, 105, 110, 115, 120, 125, 130, 135 . . . What will his weekly pocket money be on his eighth birthday, 104 weeks later? Using the formula $a + (n - 1)d$ again, the calculation is $(100 + 103 \times 5)$ pence, which come to 615 = £6.15! Maybe Mum was a bit rash. She should have done her maths first. Perhaps a 2p weekly rise might have suited her budget better.

The Sum of the Terms in an AP

The formula for **adding** the terms of an AP is:

$$S_n = \tfrac{1}{2}n\{2a + (n-1)d\}$$

How much in total did Bruce's mum pay out in pocket money for the first 104 weeks?

S_n is the total sum paid over the 104 weeks, n is the number of weeks, a is the first week's pocket money and d is the weekly increase.

$$S_{104} = 52\{2 \times 100 + 103 \times 5\}$$
$$= 52\{200 + 515\}$$
$$= 52 \times 715$$
$$= 37180$$
$$= £371.80$$

Bruce's mum shuddered to think how much the total sum would be by the time her son was a teenager, when he would expect a lot more!

The Geometric Progression (GP)

In an AP, you added or subtracted to get the next term. For example, 1, 4, 7, 10 . . . or 1, −1, −3, −5 . . .

In a GP, you multiply. For example, starting with 2 and multiplying by 3, you get the sequence 2, 6, 18, 54, 162 . . . with the starting number, a = 2, and the multiplier, called the Common Ratio (r), being 3.

Here's another sequence, with a = 4 and r = ¼:

$$4, 1, ¼, ¹⁄₁₆, ¹⁄₆₄ . . .$$

The sequence could be written out:

$$4 \quad 4 \times ¼ \quad 4 \times (¼)^2 \quad 4 \times (¼)^3 \quad 4 \times (¼)^4 . . .$$

So the general form for the terms of a GP is a, ar, ar^2, ar^3, ar^4, ar^5 . . . and again, noticing that ar^5 is actually the sixth term, the nth term will be ar^{n-1}.

Examples

(a) Find the seventh term for the GP: 4, 12, 36 . . . The seventh term is $ar^6 = 4 \times 3^6 = 2916$.

(b) Find the eighth term for the GP: 1, ½, ¼, ⅛... The eighth term is $ar^7 = 1 \times (½)^7 = ½{128}$.

The Never-ending Journey

Come with me on the never-ending journey of life...

A careless jogger trod on a tiny ant the size of a full stop and two of the insect's six legs were broken. If the poor little ant didn't get to the Ant Doctor, two metres away at the other side of the path, his days would sadly be numbered.

Dragging his poor broken legs, he set off across the path and managed to crawl one whole metre on the first day but only half a metre on the second day. Then getting weaker by the day, he managed a quarter and then an eighth, a sixteenth and a thirty-second. He was exhausted and didn't think he could carry on. He added up what had done so far:

$$1 + ½ + ¼ + ⅛ + ½{16} + ½{32} = 1\tfrac{31}{32}$$

Would he ever get there? Well, the answer is yes *and* no. No, because there is never a last fraction. They will get smaller and smaller but they will go on forever.

But, yes, because on the eighth day, the Ant Doctor spotted the poor little insect, reached out for him and carried him the very short distance to the Ant Hospital.

In maths, the ant's journey is described as: $n \rightarrow \infty$, $S_n \rightarrow 2$. Or, as the fractions approach infinity, the sum of the fractions gets nearer and nearer to 2.

There is a simple formula for calculating the sum to infinity when the terms are decreasing:

$$S_\infty = \frac{a}{1 - r}$$

where, for the ant's journey, $a = 1$ and $r = \frac{1}{2}$ so:

$$S_\infty = 1 \div (1 - \tfrac{1}{2}) = 1 \div \tfrac{1}{2} = 1 \times 2 = 2$$

There is a happy ending to the story. The fully recovered ant went back to the anthill. One day, the same jogger passed the anthill and an army of ants attacked him. Unfortunately, his running shorts were very short.

The Scary Σ

Σ is one of the symbols that convinces math-phobics that the subject is difficult and mysterious. Blame Leonhard Euler. He introduced it.

Σ is the upper-case eighteenth letter in the Greek alphabet and is called sigma. There is no mystery about it. Mathematicians use it to mean 'the sum of' or simply 'add'.

$$\sum_{n=1}^{4} n$$

simply means add the numbers (n) from 1 to 4.

$$\sum_{n=1}^{10} n^2$$

means write out as a sum of numbers, the value of n^2 from n=1 to n=10.

$$1 + 4 + 9 + 16 + 25 + 36 + 49 + 64 + 81 + 100 = 385$$

$$\sum_{n=1}^{4} 4 + 2n$$

means, when n = 1, 2, 3 and 4, 4 + 2n = 6, 8, 10, and 12. Now add them! So the answer to that little bit of mumbo-jumbo is 36.

$$\sum_{i=1}^{20} i$$

This means you simply have to add the numbers from 1 to 20, so 'do a Gauss':

$$20 \times 21 \div 2 = 210$$

More Geometry

Medians

A median is the line drawn from the vertex of a triangle to the midpoint of the opposite side. In the diagram below, L is the midpoint of BC, M is the midpoint of AC, and N is the midpoint of AB.

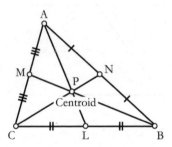

The three medians meet at point P, the centroid, and the centroid is the centre of gravity of the shape. If you are in a mathy mood, draw a triangle like the one above on cardboard and cut it out. Then place it on a sharpened pencil at the centroid and you should be able to take the triangle for a walk around the room, balancing the shape for ages. When you try this interesting little experiment, keep away from the window in case the neighbours become concerned about your sanity.

The centroid is also a point of trisection on the medians, so AP : PL = 2 : 1 as are BP : PM and CP : PN. So, for example, if AL = 3.6 cm then AP = 2.4 cm and PL = 1.2 cm.

Altitudes

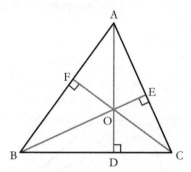

The altitude of a triangle is the perpendicular line drawn from the vertex to the opposite side. In this diagram, altitudes AD, BE and CF meet in point O, which is called the orthocentre. Unlike the centroid opposite, the orthocentre has very little use.

The Perpendicular Bisectors of a Triangle

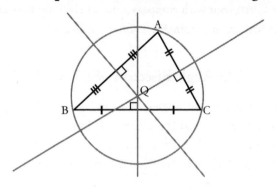

The three long lines are the perpendicular bisectors of each side of the triangle and they do not pass through the vertices of the triangle. The point Q where they meet is called the circumcentre and QA = QB = QC. With a compass point on Q and the pencil point on any of the vertices, the circumcircle can be drawn.

How to Bisect an Angle with a Compass

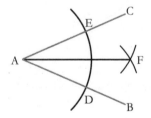

Draw any angle BAC. With your compass point at A and any

radius, draw arc ED. With your compass point at E, draw an arc, and then your with compass point at D draw another arc to cut the first arc at F. Join AF. $\angle CAF = \angle BAF$.

The Angle Bisectors of a Triangle

In the triangle below, angles A, B and C have been bisected as above. They meet at the incentre I. From I, perpendiculars are drawn to each side. Then, with centre I and radius IM, the incircle can be drawn. Admire how the circle fits snugly into the triangle.

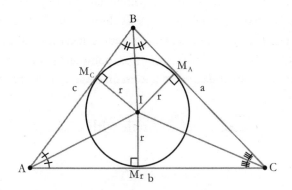

One Two Three

Choose any long number with random digits. And where better to find unlimited random digits than the digits of π. Here are the first one hundred in groups of five to save you from getting cross-eyed:

14159 26535 89793 23846 26433 83279
50288 41971 69399 37510
58209 74944 59230 78164 06286 20899
86280 34825 34211 70679

Next, sort them into even and odd digits and record them as follows:

even : odd : total
51 : 49 : 100

Write this down as a new number:

5149100

Sort into evens, odds and total again:

3 : 4 : 7

Write this as a new number and repeat the instructions:

347

1 : 2 : 3

and 123 forever.

The Same Experiment with φ

Let's try it for the first one hundred decimal digits of phi, the Golden Ratio:

1·61803 39887 49894 84820 45868 34365

63811 77203 09179 80576

28621 35448 62270 52604 62818 90244

97072 07204 18939 11374 . . .

even : odd : total

59 : 41 : 100

5941100

3 : 4 : 7

347

1 : 2 : 3

123

Try it for **any** two numbers and their total.
Whatever numbers you start with, you'll always end up with 123.

A Taste of Trigonometry

School students love trigonometry. They take to it like a dot to a decimal. Their other maths subjects are never shortened to Arith, Algie or Geom but by the end of the first lesson, they have made up their minds that they like the new subject and they call it just 'trig'.

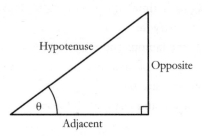

Trig is all about the measurement of the angles and sides of a triangle, and for trig the favourite symbol for a missing angle is not usually the curly x. In trig, the missing angle is often, but not always, theta (symbol θ), the eighth letter in the Greek alphabet – chosen, of course, just to give the subject a little mystery. The side across from θ is the opposite side, 'opp' or just O; the side the angle sits next to is the adjacent side, 'adj' or just A; and the hypotenuse, H, is where it always is: across from the right angle.

THE NEW
FAVOURITE
SYMBOL

The basic trig functions are sine, cosine and tangent, but they are commonly called sin, cos and tan. 'Sin' is pronounced 'sign', rhyming with 'time'. It has nothing to do with the Seven Sins, deadly or otherwise.

Now here is the famous and arguably the most useful mnemonic in the world of mathematics. It is **SOHCAHTOA.** Teenage schoolboys are particularly fond of it. They become silly little boys again and chant SOHCAHTOA, over and over again!

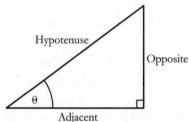

SOH means $\sin \theta = \dfrac{O}{H}$ CAH means $\cos \theta = \dfrac{A}{H}$
TOA means $\tan \theta = \dfrac{O}{A}$

209

But before we start all this exciting stuff, you will need to find sin, cos and tan on your scientific calculator:

(a) First check that your calculator is working in degrees not radians (deg not rad).
(b) Find the buttons sin, cos and tan, which you will use for finding a missing side. (Ignore sinh, cosh and tanh.)
(c) Press the second function button and find $\sin^{-1}$, $\cos^{-1}$, $\tan^{-1}$. They are the inverse functions you will use for finding a missing angle. (Ignore $\sinh^{-1}$, $\cosh^{-1}$, $\tanh^{-1}$.)

Now use your calculator to check that:

(a) $\sin 30° = 0.5$ (press 30 sin)
(b) $\sin 60° = 0.866$
(c) $\sin 45° = 0.707$
(d) $\cos 70° = 0.342$ (press 70 cos)
(e) $\cos 12° = 0.97$
(f) $\tan 45° = 1$ (press 45 tan)
(g) $\tan 80° = 5.671$
(h) $\tan^{-1}(0.577) = 30°$ (press 0.577 $\tan^{-1}$)
(i) $\sin^{-1}(0.866) = 60°$
(j) $\cos^{-1}(0.940) = 20°$
(k) Confirm $\tan^{-1}(3.43) = 73.7°$ in the chapter on gradient earlier in the book.

Solving Triangles

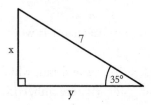

Solving a triangle means finding everything about it – all its angles and all its sides. We can do this entirely using trig or, easier, with a mixture of trig, basic geometry, (the angles of a triangle add up to 180°) and Pythagoras. Starting with trig:

This is SOH because we have the opposite side and the hypotenuse:

$$\text{Sin } 35^0 = {}^x\!/_7$$
$$0.574 = {}^x\!/_7$$
Multiply bs by 7
$$\therefore x = 4.02$$

Then by Pythagoras:

$$7^2 = 4.02^2 + y^2$$
$$49 = 16.2 + y^2$$
$$16.2 + y^2 = 49$$
$$y^2 = 49 - 16.2$$
$$= 32.8$$
$$\therefore y = 5.73$$

or you could use trig:

$$\text{Cos } 35^0 = \frac{y}{7}$$
$$0.819 = \frac{y}{7}$$
Multiply bs by 7
$$\therefore y = 5.73$$
(as above)

and lastly, the third angle is 180° − (90° + 35°) = 55°

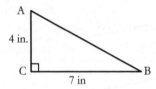

In the triangle above, we need to find one of the angles. Let's find ∠A. This is TOA because we have opposite and adjacent sides:

$$\text{Tan } A = \frac{7}{4} = 1.75$$
$$\text{So } \angle A = \tan^{-1}(1.75) = 60.3°$$
$$\angle B = 29.7° \angle s \text{ A,B, and C add to } 180°$$

And by Pythagoras:

$$AB = \sqrt{(4^2 + 7^2)} = 8.1 (1dp)$$

Or using TOA again to find ∠B:

$$\text{Tan B} = \text{}^4\!/_7 = 0.571$$
$$\text{So } \angle B = \tan^{-1}(0.571) = 29.7° \text{ (as before)}$$

In the next four triangles, we will only calculate the missing data.

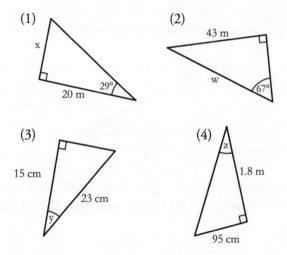

(1) This is TOA because we have the opposite and adjacent sides.

$$\text{Tan } 29° = \text{}^x\!/_{20}$$
$$0.554 = \text{}^x\!/_{20}$$
Multiply bs by 20
$$x = 11.1 \text{ m}$$

(2) This is SOH because we have the opposite side and the hypotenuse:

$$\text{Sin } 67° = {}^{43}\!/w$$
$$0.921 = {}^{43}\!/w$$
$$\text{Multiply bs by } w$$
$$0.921w = 43$$
$$w = 43 \div 0.921$$
$$= 46.7 \text{ m}$$

(3) This is CAH because we have the adjacent side and the hypotenuse:

$$\text{Cos } y = {}^{15}\!/_{23} = 0.652$$
$$\text{Cos}^{-1} (0.652) = 49.3°$$
$$y = 49.3°$$

4) This is TOA because we have the opposite and adjacent sides:

$$\text{Tan } z = {}^{95}\!/_{180} = 0.528$$
$$\text{Tan}^{-1}(0.528) = 27.8°$$
$$Z = 27.8°$$

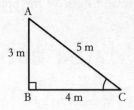

Now, all on your own, calculate the angles in the first Pythagorean triple above.

Answers: 36.9° and 53.1°

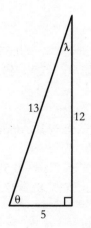

Calculate the sizes of the angles in the second Pythagorean triple above.

Answers: 67.4° and 22.6°

The Mathematics of Cutting a Pancake

Charlie, the chef at Windy Ridge Golf Club, was in a bad mood. He was overworked and underpaid and his waitress had taken the day off.

He was particularly annoyed at the lady members. He had just made a batch of dinner-plate-sized pancakes. The male members enjoyed a whole one, slathered in clotted cream and strawberry jam. But these pesky lady members always wanted to share.

'Cut it in three, please, Charlie,' said the lady captain as she sat and gossiped with her two best buddies.

Back in the kitchen, he took his chef's knife and hastily chopped through the pancake three times. Oops, big mistake! There were seven miscellaneous-sized shapes on the plate.

However, the ladies were kind. 'Goodness only knows how many bits you would have got if you had wielded that knife four times,' the lady captain laughed.

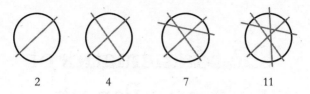

Well, that is easy to find out because pancake cutting has a formula:

$$P = \frac{n^2 + n + 2}{2}$$

where P equals the number of pieces and n is the number of cuts. For four cuts, $P = (4^2 + 4 + 2) \div 2 = 11$.

Charlie made another pancake for the ladies and carefully divided it into three 120° pieces and he was extra generous with the clotted cream.

Pascal's Triangle

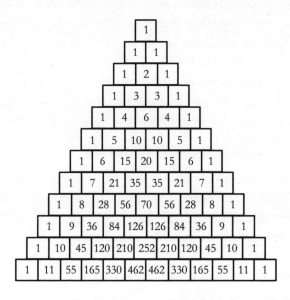

Blaise Pascal was a seventeenth-century French mathematician, scientist and philosopher, but he will be remembered forever for his famous triangle.

Like almost every mathematical pattern, only a small part of it can be illustrated on a page, but the rows of numbers will go on forever and the numbers get bigger and bigger.

The 1 at the top is there only to give the triangle a vertex and it is called Row 0. Row 1 starts with 1 1 and thereafter every row starts and ends with 1, but the other numbers are found by adding the two numbers immediately above. The number after the 1 tells you which row you are at. For example, 1, 8, 28, 56, 70. . . means you're at Row 8.

The triangle is full of mathematical gems but some are for mathematicians who are long past the need for a book like *Easy as Pi*. However, I am sure you will be amazed at the following fascinating facts.

(a) Starting at the 1 of Row 2 (first one or last one, because the triangle is symmetrical), go down the slope and there is a sequence you found early in this book: 1, 3, 6, 10, 15, 21 . . . the triangular numbers that Carl Friedrich Gauss was so fond of.

(b) Add the numbers in each row getting 1, 2, 4, 8, 16, 32 . . . or 2^0, 2^1, 2^2, 2^3, 2^4, 2^5 . . .

(c) Starting at any 1, add the numbers in the slope as far as you want. If you come down in a south-east direction, change direction to the first number south-west for the total. For example, travelling SE, 1 + 6 + 21 + 56 = (turn SW) and there is your total: 84.

Or coming down from 1 in a SW direction, add 1 + 7 + 28 + 84 + 210 = (turn SE) and read off the total, 330. This is called Hockey Stick adding because of the shape of the calculation.

(d) If, in any row, the number after the 1 is a prime number, every number in the row will be a multiple of that prime number. For example, in the triangle above, at Row 11, prime number 11 divides into 55, 165, 330 and 462. Check for 3, 5 and 7 above and if you extend the above triangle, you will find multiples of 13 and 17 all along these rows.

(e) In the triangle, pick out any hexagonal shape of six numbers. For example,

$$
\begin{array}{cc}
6 & 4 \\
10 \quad\quad 5 \\
20 \quad 15
\end{array}
$$

$$6 \times 4 \times 10 \times 5 \times 20 \times 15 = 360\,000$$
which is a perfect square ($\sqrt{360\,000} = 600$)

Let's try another one further down the triangle:

$$
\begin{array}{cc}
7 & 21 \\
8 \quad\quad 56 \\
36 \quad 84
\end{array}
$$

Now multiply $7 \times 21 \times 56 \times 84 \times 36 \times 8$
$$= 199\,148\,544$$

And would you believe it, this is a perfect square again and, just getting it to fit on my calculator, the square root of this enormous number is 14 112 exactly.

Here is the hexagon again with a triangle of numbers in bold type,

$$7 \quad \textbf{21}$$
$$\textbf{8} \qquad 56$$
$$36 \quad \textbf{84}$$

Multiply them, $21 \times 84 \times 8 = 14\,112$, and multiply the other trio, $7 \times 56 \times 36$, getting 14 112 again. Now, isn't that amazing!

Check that this works for the first hexagon and try it out on any other hexagon in the triangle.

Coin Tossing

If three coins are tossed, what is the chance of getting two heads? You can have:

$$\text{HHH} \quad \text{HHT} \quad \text{HTT} \quad \text{TTT}$$
$$\text{HTH} \quad \text{THT}$$
$$\text{THH} \quad \text{TTH}$$

So the chance of getting two heads is 3 in 8, or, from the triangle, Row 3: 1 3 3 1, giving 3 in 8.

A couple would like to have a family of four, ideally two boys and two girls, in any order.

This is Row 4 in the triangle: 1 4 6 4 1.

BBBB	BBBG	BBGG	BGGG	GGGG
	BBGB	BGBG	GBGG	
	BGBB	BGGB	GGBG	
	GBBB	GBBG	GGGB	
		GBGB		
		GGBB		

So the chance of having a perfect little family of two boys and two girls is 6 in 16 or 3 in 8.

Combinations

We can use Pascal's Triangle to calculate combinations. A combination in maths is the process of choosing r things from a set of n things where the order doesn't matter. It can be written in maths language as:

nCr

(from n choose r)

The silliest example often given is about socks. A man has seven socks in his sock drawer. How many ways can he combine two of them? Well, if he names his socks, A, B, C, D, E, F, G, he can have the following choices:

AB AC AD AE AF AG
BC BD BE BF BG
CD CE CF CG
DE DF DG
EF EG
FG

So, by adding these choices: 6 + 5 + 4 + 3 + 2 + 1 = 21 (or by doing 'a Gauss', 6 × 7 ÷ 2). The man could have discovered this more quickly by going to Pascal's Triangle, Row 7, entry 2:

1 7 **21** 35 35 21 1
Entry 2 is 21 (the 1 at the beginning is entry 0)

Only a mad mathematician would have a scant collection of seven single socks. The rest of us fold our socks together in neat pairs. And our man wasn't going to wear a red sock and a green-and-yellow-striped one – or was he? Let's find a more likely example.

Pascal's Triangle and Wagner

The principal maths teacher at the local school got himself a new car. His previous car had an unfortunate number plate. It ended in BAT, and for the last five years the kids had been calling him Batty Brown; he wanted it to stop.

His new car had built-in Bluetooth and a six-stack CD changer. He collected his eight favourite CDs, all of them Wagner, of course. Batty just loved the music of Richard Wagner and often pretended he was conducting *Ride of the Valkyries* at the Royal Albert Hall. But he would have to choose six out of eight. No problem, he numbered the CDs from 1 to 8 and decided that he would change them around by replacing a different two every Saturday morning when he washed his shiny new car. He made a list of the disks to be replaced every week:

$$(1,2) \ (1,3) \ (1,4) \ 1,5) \ (1,6) \ (1,7) \ (1,8)$$
$$(2,3) \ (2,4) \ (2,5) \ (2,6) \ (2,7) \ (2,8)$$
$$(3,4) \ (3,5) \ (3,6) \ (3,7) \ (3,8)$$
$$(4,5) \ 4,6) \ (4,7) \ (4,8)$$
$$(5,6) \ (5,7) \ (5,8)$$
$$(6,7) \ (6,8)$$
$$(7,8)$$

He did a quick 'Gauss', $7 \times 8 \div 2$, and verified that he could have a change of music every week for 28 weeks. He also consulted Pascal's Triangle, Row 8 entry 6:

$$1 \quad 8 \quad 28 \quad 56 \quad 70 \quad 56 \quad \mathbf{28} \quad 8 \quad 1$$

The kids at school admired his new car but, for some reason, they never stopped calling him Batty Brown.

A Formula for Finding Any Entry in Pascal's Triangle

There is a wonderful formula for working out combinations and any entry in Pascal's Triangle. It involves using factorials and, of course, you are now factorial experts.

$$nCr = n! \div r!(n-r)!$$

In the mini chapter about the National Lottery, you used this very formula when, for the new-style Lottery, you attempted to choose six winning numbers from a total of fifty-nine.

Let's check the sock calculation,

$$_7C_2$$
$$7! \div 2!(7-2)!$$
$$\frac{7!}{2! \times 5!}$$

$$= \frac{7 \times 6 \times 5!}{2! \times 5!}$$
$$= 21$$

225

Now you check Batty's calculation, 8! ÷ 6!(8 –6)!

You might like to know if the Lottery calculation is on Pascal's Triangle. It will be on Row 59, entry 6:

1   59   1711   32 509   455 126   5 006 386   **45 057 474** . . .

There it is! Please, please, don't expect to see the whole row! The digits would stretch from here to New York.

Combination Locks

In combination calculations, order doesn't matter. However, in a combination lock, order certainly does matter. If the code for your bicycle padlock is 762 and you try 672, the lock won't open. So combination locks have been wrongly named.

Pattern

You can make the most beautiful patterns with Pascal's Triangle, especially if you can draw them on hexagonal graph paper. Luckily, you can download hexagonal graph paper online in a variety of sizes, so have fun and get artistic.

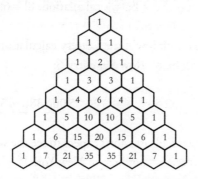

Copy the triangle above but add as many more rows as your paper will hold. Do one picture shading the odd numbers and even numbers in different colours or another pattern, colouring the multiples of 3 or 5. Be bold. Experiment.

I hope you have discovered that maths is fun.

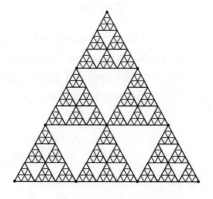

The Last Maths Joke

Question: What is the name of the vessel used for making soup at the summit of Ben Nevis?

Answer: A high-pot-in-use!

You were warned at the beginning. If you even twitched your lips in the tiniest vestige of a smile, there is no hope. You are showing the first symptoms of mathematical geekiness, for which there is no cure.

Postscript:
In Praise of Mathematics

In most sciences, one generation tears down what another has built, and what one has established, another undoes. But in mathematics, alone, each generation adds a new storey to the old structure

Hermann Hankel, German mathematician (1839–73)

ADDING A NEW STOREY

THE END

Index